BLUE SKIES OVER BEIJING

BLUE SKIES OVER BEIJING

Economic Growth and the Environment in China

MATTHEW E. KAHN AND SIQI ZHENG

PRINCETON UNIVERSITY PRESS ▪ PRINCETON AND OXFORD

The Library of Congress has cataloged the cloth edition as follows:
Names: Kahn, Matthew E., 1966– author. | Zheng, Siqi, author.
Title: Blue skies over Beijing : economic growth and the environment in China /
Matthew E. Kahn and Siqi Zheng.
Description: Princeton : Princeton University Press, [2016] | Includes
bibliographical references and index.
Identifiers: LCCN 2015045023 | ISBN 9780691169361 (hardback)
Subjects: LCSH: Economic development—Environmental aspects—China. |
Pollution—China. | Sustainable development—China. | Environmental
policy—Economic aspects—China. | China—Environmental conditions. |
BISAC: BUSINESS & ECONOMICS / Environmental Economics. | BUSINESS &
ECONOMICS / Development / Sustainable Development. | BUSINESS &
ECONOMICS / Economics / General.
Classification: LCC HC430.E5 K34 2016 | DDC 363.730951/091732—dc23 LC
record available at http://lccn.loc.gov/2015045023

British Library Cataloging-in-Publication Data is available

This book has been composed in Whitman and Helvetica Neue

Matthew E. Kahn dedicates this book to his wife Dora and his son Alexander.

Siqi Zheng dedicates this book to her parents, husband, and son.

CONTENTS

ACKNOWLEDGMENTS

The authors would like to thank the excellent team at Princeton University Press for their guidance and advice. Seth Ditchik, our editor, has provided constructive feedback and encouragement. We are also grateful to Mark Bellis and Samantha Nader for helping us with all matters of editing and design for the book, and to Brian Bendlin for his careful copyediting. We thank Joe Jackson for reading an early draft of our manuscript and for providing us with detailed comments and criticism. We also thank seminar participants at the London School of Economics and Political Science, the University of Chicago, and the University of California–Berkeley for their comments on much of the material presented in this book.

Matthew E. Kahn would like to acknowledge receiving financial support from the University of California–Los Angeles Ziman Center for Real Estate.

Siqi Zheng would like to thank Cong Sun, Antoine Nguy, Weizeng Sun, and Xiaonan Zhang for their research assistance. She gratefully acknowledges research support from the National Natural Science Foundation of China (grants 71273154 and 71322307). We would like to thank the many Chinese urbanites whom we interviewed to learn about their quality of life. Throughout the book we present quotations from several of them. To protect their privacy, we quote them using pseudonyms.

BLUE SKIES OVER BEIJING

Introduction

Mr. Wu is thirty-eight years old and has a doctoral degree in civil engineering from Tsinghua University. Originally from Hunan Province, he moved to Beijing, where he now manages a department in a large, state-owned building design company. He earns a good salary. He and his wife, who works in a state-owned hospital, have a five-year-old daughter. Like many young urban couples in China, they bought their condominium and car before their daughter was born so she could grow up comfortably.

Mr. Wu enjoys a much better quality of life than his parents did. Thirty years ago, the Communist Party of China would allocate jobs and dormitory housing to college graduates like Mr. Wu's parents. For most basic necessities—from grain, meat, and cooking oil to clothes, soap, and bicycles—the party distributed ration coupons.

But Mr. Wu faces challenges that his parents did not. If he doesn't reach his profit target at work, he faces income deductions. If he suffers from health problems, such as a serious cough from the terrible pollution in Beijing, the fierce competition with his colleagues forces him to stay on the job. When he goes out to dinner, he is careful about what he eats because he has read about the excessive levels of drugs and hormones fed to chickens as they're growing. At home, he must care for his parent in their retirement and plan his daughter's schooling. The nation's one-child policy creates extra anxiety for urban parents as they focus so much energy on their sole child's success. There is a limited number of elite slots in high-quality schools and colleges, and this puts heavy pressure on every child. Worried about his daughter's future, Mr. Wu might move to Canada or the United States.

Young Chinese also face very high home prices in the major cities. Ms. Feng has a graduate degree from Tsinghua University, and works at a major real estate company. She recognizes that the booming market has brought her company enormous business opportunities, but she laments the soaring house prices in Beijing. Her family rents a small, old apartment in the Xicheng District. To buy a hundred-square-meter condominium unit, they would have to save for ten or fifteen years. Ms. Feng describes her workweek as "five plus two, and white plus black"—that is, all five weekdays, both weekend days, and always late into the night. The extra hours are considered voluntary, so she does not receive overtime pay. Those who do not follow this routine lag behind in their performance evaluations and are pressured to leave.

Ms. Feng works a much longer week than would a typical worker in western Europe. Indeed, a comparison between daily life in urban China and western Europe yields striking contrasts. Urban China's material standard of living is rising, but urban pollution and stress are extreme. Western Europe's cities offer a high quality of life, and their inhabitants have ample leisure time to enjoy it.[1]

Over the last thirty years, China's economy has grown at an amazing rate of 10 percent per year, and the share of people living below the poverty line fell from 84 percent to 13 percent. There are still hundreds of millions of poor households in rural China, but hundreds of millions have also escaped poverty. The horrible famine of 1959–61 is now a distant memory, and improvements in medical care and diet have lengthened life expectancy. Over the last thirty years, the average life expectancy at birth has increased from sixty-six to seventy-three years.

Despite this progress, Chinese urbanites must reckon with the reality that the nation's standard of living is not improving as quickly as its economy is growing. Their cities suffer from limited access to health care and education as well as disastrous environmental quality.

The Chinese and Western media have published high-profile and lengthy exposés on environmental and other problems such as lead pollution in Deqing, toxic chemicals created by the mining of rare earths in Inner Mongolia, a proven decrease in life expectancy in northern China due to coal burning, fox and rat meat sold as mutton, and even thousands of dead pigs floating down the river in Shanghai—all salient examples of the costs of China's economic growth.

In early 2013 the incredible smog in northern China caught the world's attention.[2] In January 2013 the particulate matter concentration in Beijing reached levels of two, three, and even four times the public health emergency threshold of 250 micrograms per cubic meter—and up to forty times what the World Health Organization (WHO) considers a healthy level.[3] Based on one key indicator of outdoor air pollution, twelve of the twenty most polluted cities in the world are in China.[4] In 2003, 53 percent of the 341 monitored Chinese cities—accounting for 58 percent of the country's urban population—reported annual average pollution levels that exceeded the WHO's standard. One percent of China's urban population lives in cities that meet the European Union's air-quality standards.[5] One study estimates that such extreme pollution may cause twelve hundred premature deaths per year in Hong Kong alone.[6]

Another cause for concern is water pollution. According to a report by the Chinese Ministry of Environmental Protection, 57 percent of the groundwater in 198 cities was officially rated as "bad" or "extremely bad" in 2012, while more than 30 percent of the country's major rivers were found to be "polluted" or "seriously polluted."[7]

China is the world's largest emitter of greenhouse gases, and these emissions exacerbate the risk of climate change. While per capita energy consumption in China is still less than 30 percent of that in the United States, China's total energy consumption surpassed total US energy consumption in 2009. Data from the World Bank shows that China's per capita greenhouse gas emissions grew by 186 percent (to 5.2 tons per person) between 1990 and 2010, while the world's emissions grew by 16 percent (to 4.9).[8]

We've Been There

Today China faces many local environmental challenges, and an unintended consequence of its industrial production, increased motor vehicle use, and coal reliance is growing greenhouse gas emissions. In contrast, cities in the United States have enjoyed great progress toward cleaner air and water in the last forty years.

Not long ago, the cities of the West were much more polluted. Coal burning in major cities such as London and New York City created soot that killed thousands; London's Great Smog of 1952 alone killed at least four thousand (and by some estimates, as many as twelve thousand) people as coal emissions from residential burning greatly elevated local particulate levels. Also in the mid-twentieth century, heavy manufacturing in major cities like Los Angeles, New York, and Pittsburgh (whose booming steel industry offered high-paying but dirty jobs) led to severe air and water pollution. Meanwhile, rising motor vehicle use relying on leaded gasoline caused high levels of urban lead emissions. In the 1960s and 1970s, smog in Los Angeles increased dramatically due to an increased number of vehicles traveling greater distances.

But the combination of new regulations (perhaps spurred by the horrible consequences of the Great Smog), energy efficiency gains, and rising household incomes that encouraged the substitution away from dirty fuels such as coal toward cleaner fuels such as natural gas fostered air-quality improvements for dense cities during times of growth. The birth of the environmental movement in the 1960s, often associated with the publication of Rachel Carson's *Silent Spring*, helped mobilize the growing number of educated people in US cities to work toward a cleaner environment and preserving natural capital.

In the case of vehicle emissions, effective environmental regulations in the United States have offset the growth in total annual miles driven. Vehicles built in 2015 emit 99 percent less local air pollution per mile than those built before 1975. Thus, despite continuous vehicle use growth and increased mileage, over the last several decades levels of Los Angeles smog have plummeted.[9]

Starting in the early 1960s, Pittsburgh and other Rust Belt cities lost thousands of manufacturing jobs. The silver lining was blue skies: as industrial activity declined, air and water quality sharply improved. Pittsburgh reinvented itself as an attractive city on the Allegheny and Monongahela rivers, with new firms that relied on an educated workforce benefiting from access to leading research universities such as Carnegie Mellon and the University of Pittsburgh. Boston and Chicago enjoyed similar transitions to blue skies, as did London.

The US experience offers some lessons for predicting future dynamics for China's environmental quality. While the two nations differ on many levels, the experience of cities in the United States highlights the role that fossil fuel consumption, the scale of industrial activity, and private vehicle use play in contributing to urban pollution. A city of given population size will experience an improvement in environmental quality if its power plants and industrial boilers move away from coal, there is a transition away from heavy industries, and firms introduce new technologies that reduce emissions. The transportation sector will create fewer emissions if people drive less or if private vehicles emit less pollution per mile of driving. The two key variables here are the scale of economic activity (i.e., industrial production or total miles driven) and the pollution intensity per unit of economic activity. For a growing economy to accomplish improvement to the environment, pollution per unit of economic activity must decline faster than economic activity grows. For example, if the people of Beijing drive 100 percent more miles in the year 2015 than they did in the year 1990, aggregate vehicle emissions can only decline if emissions per mile of driving decline by more than 50 percent over this same time period. Tracing the scale and the pollution intensity of economic activity in a growing city provides a framework both for tracking pollution dynamics in a Chinese city and comparing Chinese cities' environmental performance over time.

Reasons for Hope

Will the 2013 Beijing haze be China's equivalent of the Great Smog of 1952—a catalyst for genuine environmental change?[10] There are several trends now unfolding in China that suggest that many of China's cities will experience positive environmental change in the coming decades.

Of eighty-three major Chinese cities for which we can access urban air pollution data (measured as particulate matter up to ten micrometers in size, or PM_{10}) we predict that forty-nine will experience near-term progress in curbing air pollution.

From 2001 to 2013, Beijing's annual ambient particulate levels have declined by 39 percent. This reduction in pollution has taken place at a time when Beijing's population, number of motor vehicles, and per capita income have continued to grow. An examination of PM_{10} levels across eighty-three of China's major cities over the years 2005–10 indicates that, controlling for a city's population size and its share of employment from manufacturing, pollution is declining by 2.8 percent per year. Assuming that this past statistical relationship continues to hold, we predict that a city whose population and manufacturing share does not change over time would enjoy a 28 percent decline in PM_{10} levels over a ten-year period. While China's cities are growing in size, and city size is positively correlated with PM_{10} levels, the impact of city growth on urban pollution levels is small. A 10 percent increase in a city's population (e.g., Beijing growing by two million people) is associated with only a 1.3 percent increase in ambient PM_{10} levels.

Many Chinese urbanites are becoming increasingly aware of the threats and impositions on their quality of life, and as more people obtain higher education and better wages, their standards and demands are rising. Indeed, the pollution in Chinese cities has sparked widespread complaints and calls for a cleanup.[11] Via the Internet, the Chinese people are discussing and debating the causes and consequences of urban pollution. *Under the Dome* is a 2015 self-financed Chinese documentary film produced by Chai Jing, a former China Central Television journalist. The film, which openly criticizes state-owned energy companies, steel producers, and coal factories that are responsible for pollution, has struck a nerve in China; within three days of its release it was viewed over 150 million times on the Tencent video portal. Chen Jining, the former president of Tsinghua University and, as of February 2015, the head of China's Ministry of Environmental Protection, praised the film, comparing its significance with Carson's *Silent Spring*.[12] Over 18 percent of the world's population lives in China, and a majority of China's population now lives in cities. The quality of life for the growing urban middle class is a key determinant of political stability for the nation, the region, and the world. [13]

A Preview

In this volume we seek to understand how China's urban economic growth impacts local and global environmental challenges, and we adopt a microeconomics perspective focused on the choices made by households, firms, and various levels of the Chinese government that in aggregate impact the environment. No rational actor actively seeks to damage the environment; instead, environmental damage often emerges as an unintended by-product of individuals' choices and firms' production decisions.

To understand how improvements in the environment could take place, we must identify the incentives that would allow Chinese cities to achieve improved environmental performance. For example, why do Chinese industrial plants use coal if burning this fuel causes so much pollution? The economic approach asks who bears the costs and who gains the benefits from such a practice. If there are social costs associated with coal burning (i.e., hazards that a factory brings to bear on the surrounding residential area), do any local government officials have an incentive to protect the residential communities, or are these officials close to the polluting firms and thus hesitant to regulate them? If such factories unintentionally elevate local air pollution, what self-protection strategies can Chinese urbanites use to protect themselves?

We have been working together on joint research projects related to China's urban development and pollution challenges since 2006. Over the years, Matthew E. Kahn has visited and lectured in China, and Siqi Zheng has been a visiting scholar at various US research universities. This international collaboration has allowed us both to more clearly see the strengths and weaknesses of our respective political and economic systems. This book is stronger than if either of us had tried to write it alone, as it yields a more balanced examination of the challenges and opportunities for China's cities.

Matt is an expert in environmental economics, a branch of applied microeconomics that seeks to understand the causes and consequences of pollution production. Siqi is an expert on the Chinese urban economy and real estate markets. In writing this book together, we seek to convey our excitement about our joint research discoveries and to bring our research to life by weaving in personal stories about life in modern urban China. Such personal observations of people like Mr. Wu and Ms. Feng allow us to explore the human element of the massive urbanization that is now unfolding.

We seek to understand the emerging quality-of-life challenges in China from a microeconomic perspective. For China's hundreds of millions of urbanites, how does pollution affect their daily quality of life? How do their day-to-day choices in aggregate impact local and global environmental challenges? Why is their demand for a cleaner environment likely to increase over time? How will government policies influence urban environmental quality dynamics?

In chapter 2, we study the scale and the economic geography of China's massive urban industrialization. As China produces an increasing number of goods, using more electricity generated from coal, a tremendous source of pollution is its energy-intensive manufacturing sector. In 2013, 67.5 percent of China's energy was fueled by coal, compared to 20.1 percent in the United States.[14] Indeed, the total amount of China's coal production is almost equal to that of the rest of the world's nations combined. And the demand for electricity will only rise as China's urbanites grow in number and wealth. We devote careful attention to how this growth impacts quality of life and local and global environmental challenges.

Beijing, Hong Kong, and Shanghai are well known cities, but China has hundreds of other cities scattered across its 3.7 million square miles. More than one hundred of those cities has a population of over one million, and China's urban growth is just beginning; over the next thirty years, 300 million people are expected to move to China's cities. As cities take on hundreds of millions more inhabitants, the way in which they grow is changing, as is how people locate themselves and move around within the urban environment; these changes have important environmental consequences.

The Chinese government has begun to relax its internal passport system, allowing people more freedom to choose where to live. Until recently, citizens who did not have an internal passport to live and work in a specific city were denied access to that city's schools and had less access to local hospitals and retirement pensions. In late July 2014, however, China's central government announced its intention to reform the *hukou* system of internal passports. According to the announcement, "the government will remove the limits on *hukou* registration in townships and small cities, relax restrictions in medium-sized cities, and set qualifications for registration in big cities."[15] While China's leaders are not chosen by popular vote, the public will have the opportunity to reveal their preferences by where they choose to live and work—they will vote with their feet. In chapter 3, we discuss China's system of cities and present facts about the quality of life in different Chinese cities. These details provide readers with a sense of the options that Chinese urbanites have to choose from.

Environmental challenges vary both across cities and within cities. An old saying is that the "solution to pollution is dilution." Recognizing this point, many people in the United States choose to live in suburbs, where population density is lower. Suburban living offers ample green space and cleaner air, but environmentalists counter that this lifestyle raises a household's total carbon footprint because its members drive more and use more residential electricity than they would if lived in a city. In chapter 4, we examine the trade-offs of center city versus suburban living in urban China and discuss the aggregate environmental consequences of such choices. In the United States, most growth until recently has been at the suburban fringe. Chinese cities are much denser than their US counterparts, and 80 percent of big-city households live in high-rise condominium buildings. But as China's urbanites grow wealthier, will they embrace the American way of living and working in suburbia? What are the environmental implications if such a lifestyle takes root, especially if this means that more people will be driving?

In 2010, 15.2 million new vehicles were registered in China. In a nation with rising per capita income, more people are seeking the private mobility that US urbanites take for granted. The rise of motor vehicle use in China increases local air pollution and global greenhouse gas emissions from fossil fuels. In chapter 5 we discuss emerging microeconomic trends with respect to private vehicle use in China and analyze the environmental impacts of these trends; we survey the academic literature that examines how to design effective public policies to curb the local and global environmental externalities associated with such driving.

Together, our examination of the economic geography of industrial production, population, vehicle use, and the evolving urban form of Chinese cities allows us to explore the likely dynamics of pollution across and within China's cities. By understanding the microeconomics of pollution production and its spatial distribution, we begin to see the opportunities for achieving significant environmental improvements.

Once we understand what's happening with pollution in Chinese cities, we explore the demand for a cleaner environment, using market data and new survey results. As economists, we are quite interested in how people express their priorities in avoiding pollution and environmental risk based on how they "vote with their wallets." In 2008, for instance, in response to the reported presence of tainted milk, parents in major Chinese cities bought more than 60 percent of their milk for infants from overseas, and they paid a 33 percent price premium for it. Through this and other examples, in chapters 6 and 7 we study how the demand for reduced risk and pollution has increased over time in China's cities as urbanites have become better educated and wealthier. An examination of a day in Siqi's life in Beijing provides a revealing glimpse into how urban air pollution impacts her on a daily basis, as well as the steps she takes to protect herself and her family from the urban hazards they face.

A Day in Siqi's Life

Today is a typical winter Monday in Beijing; the temperature in the morning is always below freezing. Siqi wakes up very early and takes a look at the air pollution monitor application on her iPhone, which reports two versions of the city's air pollution index: one from the US embassy in Beijing, and the other from China's Ministry of Environmental Protection (MEP). It will be a terrible day again: the US embassy index reports a "hazardous" day, and the MEP reports that the air is "highly polluted."

Like other successful Beijing urbanites, Siqi has several protection strategies to reduce her exposure to pollution. She owns a car, and on days with heavy pollution she drives to work rather than riding her bike. There are many products on the market to protect people from air pollution; an air purifier costs US$490, and an air mask costs ninety cents. Each mask, which researchers believe reduces one's exposure to pollution by 33 percent, is effective for ten days.

Wearing a mask isn't glamorous, but exposure to thirty minutes of outdoor air on a hazy day in Beijing causes a sore throat. On highly polluted days, most people walking or riding bicycles wear masks, and Beijing's supermarkets, pharmacies, and shops are often sold out of them—especially the higher-quality 3M Particulate N95 masks, which are 88.5 percent effective in reducing exposure to the smaller $PM_{2.5}$ particles.

Siqi decides to drive her car today to protect herself from the polluted air. Her commute to Tsinghua University can take either twenty minutes by bicycle or ten minutes by car without traffic. During peak hours the trip by car takes thirty minutes, and in very bad traffic it can take an hour. To avoid traffic, she often gets to her office before 7:00 a.m. Thanks to the flexible working hours that professors enjoy, she can adjust her commute time to avoid traffic, but many of her friends aren't that lucky. Buses and subways are extremely crowded during rush hours, and the upper middle class and wealthy avoid public transit, choosing instead to drive their cars even in severe traffic. The average one-way commute in Beijing takes fifty minutes, which is much longer than the average commute in Los Angeles or New York City.

During the winter months, centralized heating warms Siqi's office to a comfortable temperature. Generating this heat contributes to urban air pollution in Beijing, as coal is the major source of energy. With centralized heating, people cannot control the indoor temperature, and sometimes it is so hot indoors that they have to open windows, leading to wasted energy and unnecessary pollution. Professors at Tsinghua University's School of the Environment and Department of Building Technology are developing a decentralized heating-control technique that will allow households to adjust the heat inflow and their indoor temperature independently. A few new residential complexes have adopted this technology, and it promises significant energy savings.

Siqi's condominium unit is 108 square meters in area, which is much smaller than the typical single-family house in the United States. It has two bedrooms, one living room, and one bathroom. Her condominium is located in a residential complex called Qing Feng Hua Jing. The complex has six high-rise residential towers that contain about one thousand housing units; it has a feel similar to that of Peter Cooper Village in Manhattan. There is some green space in the middle of the towers, where Siqi's son can run around with his friends. It is a typical residential community for most of the middle class in Chinese cities.

Siqi's husband works for a real estate development company in Beijing's central business district. He has a long daily commute that takes an hour by car each way. Sometimes he takes the subway, but the closest subway stop is a long walk from the condominium, and after he gets off the subway it can be very difficult to find a cab, especially when there is rain or snow. He arrives home after 7:00 p.m.

Siqi's parents are in their seventies. When Siqi had her son four and a half years ago, her parents came to Beijing, and they have been living with her and her husband ever since. On days with low pollution levels they walk around the community in the morning and they take their grandson outside to play in the community yard during the day. He makes friends there, and Siqi's parents make friends with other children's grandparents. During hazy days, they avoid their morning exercise and don't let their grandson go outside, as the local newspaper advises old people, children, and those with health conditions to stay indoors. Siqi's husband bought an indoor air purifier and turns it on for the whole day, hoping to reduce the effect of the bad air on the family.

Siqi and her family look forward to the weekends, when they like to have lunch outside. Sometimes they drive several hours to visit rural areas or small towns for fun. On one weekend outing in the fall, they might drive to a small village to pick cherries at one of the small businesses run by villagers who raise apple trees, cherry trees, and vegetables; during the growing season, people from big cities come and pay a fee to pick fruit.

Despite our narrow focus on a day in Siqi's life, these stories about her generalize to some degree. All residents of Beijing face the challenge of high home prices, air pollution, and traffic. Siqi's high level of education and her occupation have provided her with the resources and the flexibility to cope with many of these quality-of-life challenges; those who are poorer or less-educated have fewer coping strategies. But money alone doesn't shield China's urban elite from stress, pollution, and traffic.

How Will the Government Respond to the Pollution Challenge?

The rise of social media, through the microblog Weibo (the Chinese version of Twitter) and through Weixin (WeChat), has enabled Chinese urbanites to express their concerns about quality of life. As city dwellers become wealthier and more educated, they are increasingly likely to value safety and green space and thus their demand for information and political accountability will also rise.

This creates incentives for the media to cover such stories, and as a result it is playing a more active role in calling for accountability and transparency. On a foggy day in October 2011, the US embassy's particulate-matter reading ($PM_{2.5}$) was so high compared with the standards set by the US Environmental Protection Agency that it was listed as "beyond index." China's own assessment, however—based on a different standard that measured only larger particles (PM_{10})—was merely "slightly polluted." This large discrepancy was evident for several days and triggered a prominent debate in the media and on microblogs. At first Beijing officials argued that these two readings measured different-size particulates so that a direct comparison was not valid, but the public was not convinced. Later the central government stepped in and required municipalities and cities to monitor and report $PM_{2.5}$.

Will Chinese national and local governments be up to the job of addressing these concerns? The environmental degradation of Chinese cities owes much to the poor political accountability of local officials. In choosing which local officials to promote, China's central government has traditionally focused on economic growth as the main performance criterion. Local officials boosted their economies through encouraging the growth of dirty industries, and had little incentive to reduce energy consumption or protect their jurisdiction's environment.

There are several trends that encourage the Chinese central government to shift its focus. Soaring energy consumption and electricity shortages have raised domestic energy security concerns. And as the rest of the world embraces a low-carbon energy agenda, China has an incentive to carve out new export markets if it can become a technological leader in this nascent field. The government also seeks legitimacy with the Chinese people and in the international arena; by pursuing environmental improvements, it credibly signals to the international community and to the Chinese people that the Communist Party leadership cares about the quality of life of its people. Indeed, at the turn of the twenty-first century, China's central government has been gradually placing more emphasis on energy efficiency and quality-of-life targets, and we discuss these in detail in chapter 8. The Chinese government's tenth (2001–5), eleventh (2006–10), and twelfth (2011–15) five-year plans have all set specific energy-efficiency and pollution-reduction targets, and those targets are also included in local officials' promotional criteria.[16] China has a strong, one-party, central government. The State Council appoints the governors of provinces, municipalities, and some major cities (so-called provincial-level and vice-provincial cities) directly. Provincial governments appoint the governors of prefecture-level cities. The selection and promotion process is performed by the upper level of the Chinese Communist Party Committee's personnel department. Some data used in the promotion process include performance evaluations with objective and quantitative targets, individual interviews, and qualitative assessments of capacity and potential.[17]

Environmental issues have been added to the policy agenda of many city leaders, and they have tried to meet those targets, but skeptics point out that such a rigid target-based system raises verification challenges. The new rules will be more likely to yield environmental improvements only if local leaders view such improvements as beneficial for themselves,[18] and corruption and side deals between factories and local politicians could slow any progress. Officials sometimes have a direct financial stake in factories, or personal relationships with their owners, and such relationships provide financing and connections for private firms. Since 2013 Chinese president Xi Jinping has punished officials for engaging in such corrupt practices. From the local officials' perspective, a key issue arises concerning what they perceive to be their city's "golden goose." In the past, the answer was heavy industry, but today the prospects for improvements in environmental performance are enhanced if local leaders increasingly prioritize high-tech industries and the service and tourism sector.

Local leaders will be more likely to pursue a green agenda if such investments dovetail with their vision for their cities' future; for example, a mayor who wants to increase local tourism has strong incentives to keep the local beaches clean. In the language of economics, there are certain policies and regulations that are "incentive compatible." In 2007, three researchers who study this concept were awarded the Nobel Prize in economics; their research provides clues about how the Chinese central government can devise the "rules of the game" to incentivize local leaders to invest more effort in protecting the environment even though these leaders have considerable discretion with respect to their day-to-day choices.

In chapter 9 we explore the key idea that self-interested local leaders are more likely to act in the public interest if they believe that citizens can hold them accountable. A key issue is whether Chinese national and local government officials have strong incentives to address green challenges, and it is something we explore in detail.

China's urban growth has contributed significantly to global climate change. In 2012, China produced 27 percent of global carbon dioxide emissions, largely due to rising coal and gasoline consumption. In chapter 10 we explore how China's urban growth has affected its greenhouse gas production and use official statistics and the peer-reviewed literature to offer a preview of possible carbon production dynamics out to the year 2040. These results highlight that the amazing scale of China's growth will continue to increase the nation's greenhouse gas emissions unless there is significant technological change through the increased use of electric vehicles and less-expensive, renewable sources of power such as wind and solar systems. The Chinese central government has announced ambitious goals to increase the economy's energy efficiency. Whether these goals can be achieved hinges on a variety of microeconomic variables that we will discuss in the next section.

A Possible Path for Sustainable Growth

Labor economists emphasize that human capital and the development of the next generation of productive adults is the key to sustainable urban economic growth. In this book we document that much of China's past urban growth has relied on the smokestack model, an approach that yielded economic growth but also led to very high levels of local and global pollution.

At the time of this writing, in 2015, we can already observe many coastal Chinese cities experiencing a drop in air pollution. In this book we present new evidence on how many Chinese cities, and how many total urbanites, are likely to enjoy environmental improvements in the short term. Our findings build on the environmental Kuznets curve (EKC) hypothesis that suggests that an inverse-U association exists between per capita income and pollution.[19] Intuitively, this hypothesis posits that as poor cities grow

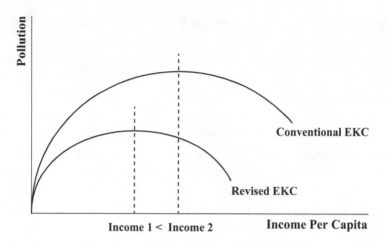

Figure 1.1 The environmental Kuznets curve: different scenarios

wealthier, such economic growth causes environmental degradation, but that as middle-income cities grow wealthier, such economic growth contributes to environmental improvements. In chapter 10 we study the intercity relationship between air pollution (measured as PM_{10}) and per capita income. We estimate that those Chinese cities whose per capita gross domestic product is greater than US$13,000 are past the turning point, so that economic growth is positively associated with improvements to the environment.

Our explanation for this statistical fact is based on several emerging trends. Chinese urbanites who reach such levels of income are increasingly well educated, and they have traveled outside of China (e.g., to Japan and the United States). Given that many adults have only one child, they seek a safe environment to raise the likelihood that that child grows up happy and healthy.

Due to China's importing world technology, China's cities can more cheaply achieve pollution emissions reduction. Figure 1.1 suggests an optimistic hypothesis that the EKC curve shifts down and in over time as the adoption of cleaner technology for cars, power generation, and winter heating means that the same level of economic production and consumption causes less pollution. We believe that this hypothesis that the EKC is shifting down and in is likely to hold true in China and to be especially likely in those cities with younger, more educated mayors and those cities where the people are more interested in environmental issues.

Our city-level results suggest that thirty-three out of these eighty-five cities in China (where 140 million people live) are already experiencing a lowering of PM_{10}. For a decrease in pollution to take place during a time of growth means that emissions per unit of economic activity must be declining in the near future. If this inverse-U pattern exists, it will be expected that, with rising income, twenty-three of the remaining fifty-two cities will pass the income turning point by the year 2020. This means that 368 million urbanites will enjoy better air quality. The average city in this group is predicted to enjoy an annual reduction in PM_{10} of 3.4 percent over the years 2012–20. The top five cities with the largest PM_{10} declines are Dalian, Guangzhou, Ningbo, Suzhou, and Xiamen.

When people are healthy and happy they are more productive workers. Environmental quality is an essential input in producing aesthetically pleasing cities and healthy inhabitants. In this sense, blue skies and a productive, happy workforce go hand in hand. Educated urbanites will be willing to pay for products and housing in neighborhoods that offer a higher quality of life and fewer environmental and health risks. Firms will cater to this demand by designing products and neighborhoods that deliver higher-quality experiences. Urban governments will be incentivized to deliver green cities both to comply with national government objectives and to meet the demands of constituents. China's citizens and leaders are now increasingly aware that a by-product of their economic growth miracle is horrendous pollution.

Countering this optimism, however, are legitimate concerns about the global effects of China's growth. Many grant that China will take steps to protect its own public health but ask about the nation's incentive to address global challenges. China's reliance on burning coal to generate electricity and for winter heating has had substantial global environmental implications. An unintended consequence of China's current energy choices is extremely high levels of local urban pollution. The health and quality-of-life costs of such particulate pollution are borne by Chinese urbanites, and this provides a direct incentive for the people and their leaders to seek out policy solutions.

To narrow this book's focus, we chiefly study how China's urban growth impacts local environmental issues and how the evolution of local pollution challenges impacts urban growth. In turn, we investigate how China's urban growth impacts its greenhouse gas production. This is the only global environmental indicator that we examine, though we recognize that there are other relevant environmental criteria, such as mineral extraction from African nations and the overfishing of the world's oceans; as China grows wealthier, its urbanites' demand for such imports could certainly exacerbate environmental challenges. That said, our focus is on urban quality of life in China, and climate change risk (in part exacerbated by China's own greenhouse gas emissions) has a direct impact on that quality of life.

We now turn to the rise of China as an industrial urban economy. Heavy industry and coal-fired generation of electricity has been the economy's engine but, as we'll see, it has also been the main cause of China's pollution problems.

PART I

A Geographic Overview of Urban Pollution Production in China

CHAPTER TWO

Made in China

In Yiwu, a small city in the Yangtze River delta in eastern China, over seven hundred manufacturers produce Christmas trees, tinsel, toys, and statuettes of Santa Claus himself. Roughly 80 percent of the world's Christmas products are made in China, and more than 60 percent of that is produced in Zhejiang Province, where Yiwu is located.

"Made in China" is a phrase that is ubiquitous in US homes, where countless products, from televisions and appliances to clothes and toys, and even pet foods, are stamped with it. The computers used to write this book are likely to have been manufactured in the Guangdong Province, and the flat-screen TV in your living room is likely to have been built in the Jiangsu Province. For decades China's manufacturing plants have had a cost advantage because labor is less expensive in China; products cost less to produce, and Walmart and other leading retailers fill their shelves with Chinese products.

China's comparative advantage in producing manufactured goods is based on the low wages it pays workers and the strong central government's policies intended to use industrialization to pull the nation out of poverty. Industrial electricity is also subsidized,[1] and it is even lower in industrial parks to attract foreign direct investment.[2] But workers' wages and energy represent only monetary costs of production; there are additional social costs associated with the pollution generated by industrial production. In 2011 China's manufacturing sector consumed about 70.82 percent of the country's total energy, while in the United States, manufacturing's share reached a peak of 41 percent in 1951, and it was only 32 percent in 2013.[3]

In this chapter, we first discuss the macrolevel environmental challenges that China's remarkable manufacturing growth has caused; coal burning is a major source of China's pollution. We then focus on the economic geography of trends in industrial production. For example, some coastal cities are deindustrializing while more inland cities are industrializing. We seek to understand what these trends mean for local quality of life. We end the chapter by discussing several nascent trends that create the possibility of reductions in the local pollution and greenhouse gases that are associated with industrial production.

The Environmental Impact of China's
Aggregate Industrial Production

Environmental economists have developed a simple accounting framework for measuring sustainability trends. Consider a manufacturing industry such as steel; to judge its environmental impact, we need to know its annual scale of production. For example, a steel manufacturer produces a given tonnage of steel; we then need an estimate of the firm's energy intensity, which represents how much energy is used to produce one ton of steel. Within the same industry, firms may differ with respect to their energy intensity because the technology built into their factories and their management's quality and efficiency differ. Each individual factory's energy consumption can be broken down into energy consumption per unit of production (technique) multiplied by its scale of production (scale). Adding up this consumption across all firms in the industry yields the industry's total energy consumption. In an economy such as China's there are many different industries that differ with respect to their energy intensity. Industries such as steel require a large amount of energy per dollar of output while other industries such as apparel use much less energy per unit of output. These cross-industry differences in energy intensity represent a composition effect. Using the accounting framework based on breaking down total emissions into scale, composition, and technique effects, we now provide some facts about the environmental impacts of China's industrial growth. Over time, the industrial sector's environmental impact can increase if there is more output (scale), concentrated in

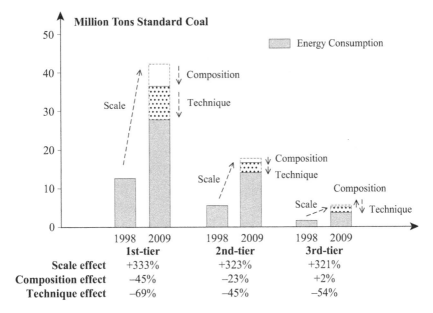

FIGURE 2.1 Industrial energy consumption changes
in Chinese cities, 1998–2009

Note: Energy consumption is measured in tons of standard coal.

Source: International Energy Agency, http://www.iea.org/; authors' calculations.

energy-intensive industries (composition), and if these industries are becoming less energy efficient over time (technique).

In figure 2.1, we report the change in energy consumption (measured
in tons of standard coal) between 1998 and 2009 for three groups of Chinese cities: first-tier cities (Beijing, Guangzhou, Shanghai, and Shenzhen); second-tier cities (all other provincial capital cities), and third-tier
cities (all other prefecture-level cities). Among 287 prefecture-level cities, 4 of them are classified as first-tier cities, 31 as second-tier cities,
and 252 third-tier cities. We break down this energy consumption change
into three components: scale, composition, and technique effects. *Scale
effect* refers to the absolute growth of a city's industrial activity; *composition effect* refers to shifts in a city's industrial structure, such as the service
sector's share of total economic activity and the shrinkage of the manufacturing sector; and *technique effect* refers to a specific industry's energy
intensity, which could decline over time due to technological progress or

environmental regulation. These three effects capture the intuitive idea that a city could enjoy a cleaner environment if overall economic activity declined (scale), if the activity concentrates in cleaner industries (composition), or if specific industries experience a reduction in pollution or energy intensity per unit of output created (technique).

Figure 2.1 highlights that, for all three city groups, the scale effect always contributes to an increase in energy consumption, but the technique effect always leads to the reduction of energy consumption. The composition effect is small. The rising industrial energy consumption in the first-tier cities over the past decade is mainly caused by the scale effect (a 333 percent increase). The composition and technique effects help reduce energy consumption levels.

Coal Burning and Electricity Generation

If China produced its electricity using nuclear and renewable power, there would be fewer local or global environmental impacts from its amazing manufacturing growth. But China currently produces roughly 80 percent of its power using coal. Such reliance on coal has horrible environmental implications for its citizens and toward exacerbating the global climate change. As we mentioned in chapter 1, the total amount of China's coal production is almost equal to that of the rest of the world combined (see fig. 2.2).

There are thousands of power plants located near Chinese cities. Nearly half of these coal-fueled plants are built close to large- or medium-size cities with five million or more inhabitants, and most of those plants are located in northern China, which includes Shanxi Province and Inner Mongolia, areas that are endowed with abundant coal resources. Taiyuan, the capital city of Shanxi Province, is one of China's dirtiest cities, and it has suffered from high sulfur dioxide and particulate matter up to ten micrometers in size (PM_{10}) concentrations since the 1990s.

Table 2.1 shows the sources of electricity generation in China and the United States in 2010. The corresponding carbon dioxide and sulfur dioxide emission factors of those fuel sources are from US Emissions and Generation Resource Integrated Database data. China relies on coal to generate

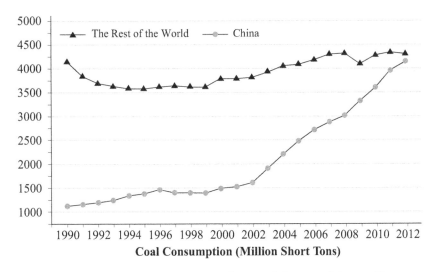

FIGURE 2.2 Coal consumption in China and the rest of the world

79 percent of its power, while for the United States the coal share is 46 percent. Natural gas is much cleaner, having a carbon dioxide emissions factor that is 42 percent that of coal and a much lower sulfur dioxide emissions rate; it is used to generate 23 percent of US electricity but only 1.4 percent of power in China. We use the US emissions factors reported in table 2.1 and combine them with annual data for China and the United States on the share of power generated from different sources such as coal, natural gas, oil, and renewables in order to calculate each nation's annual average emissions factor. China's emissions per unit of power for both carbon dioxide and sulfur dioxide are much higher than the United States because of China's reliance on coal.

Cong Sun is one of Siqi's doctoral students. The largest state-owned coal-fired power plant was built in his hometown in 1959. When Cong was a child in the early 1990s, he rode his bicycle on the street near the plant and the black dust fell on his clothes and arms. A massive amount of coal is shipped to this plant by industrial railway every day, and the old plant is still one of the city's major sources of electricity (see fig. 2.3).

One team of economists estimated that in the United States coal-fired power plants are responsible for 25 percent of total industrial pollution

TABLE 2.1 Electricity generation by sources, and corresponding CO_2 and SO_2 emission factors

Electricity generation by sources	CO_2 emission factor (pounds per MWH of electricity)	SO_2 emission factor (pounds per MWH of electricity)	China: 2010 source shares	US: 2010 source shares
Coal	2204.73	7.422	79.10%	45.80%
Natural gas	932.50	0.043	1.36%	23.38%
Oil	1700.23	5.755	0.44%	1.10%
Other green sources (water, nuclear, wind, etc.)	0	0	19.10%	29.72%

Sources: The average CO_2 and SO_2 emission factors for 2010 are calculated by using data from the US Environmental Protection Agency's Emissions and Generation Resource Integrated Database, http://www2.epa.gov/energy/egrid; the source shares for 2010 are collected from World Bank indicators, http://data.worldbank.org/indicator, and include electricity production from coal sources (% of total), electricity production from natural gas sources (% of total), and electricity production from oil sources (% of total)

FIGURE 2.3 A coal-fired power plant in the city of Jilin

damage, or $53 billion of environmental damage each year.[4] In recent years, the United States has reduced the share of electricity generated from coal and has increased its reliance on natural gas. This shift is likely to offer significant local air-quality benefits as ambient sulfur dioxide and oxides of nitrogen decline.

Coal burning releases large quantities of polycyclic aromatic hydrocarbons (PAHs) and other pollutants; PAHs are reproductive and developmental toxicants, mutagens, and carcinogens. Looking into the effects of coal pollution in China, a research team documented what scientists and economists call a "natural experiment" in the city of Chongqing's Tongliang District. A natural experiment is a case in which a random, specific event—such as a volcano erupting or an unexpected change in the weather—provides researchers with an opportunity to conduct a before-and-after comparison to attempt to establish a causal effect. In this case, the event was the closing of a coal-fired power plant; when it shut down in May 2004, the variation in coal-fired plant activity provided an opportunity to study the health outcomes for two identical groups of nonsmoking women and their newborns in 2002 (before shutdown) and 2005 (after shutdown). Their findings confirm that neurobehavioral development in Tongliang children benefited from the elimination of PAH exposure from the coal-burning plant.[5]

Another study estimated that the 500 million Chinese who live north of the Huai River lost a combined 2.5 billion years of life expectancy due to the extensive use of coal to power boilers for heating throughout the region. Under lingering communist policy, cities just north of the river received free winter heating while sister cities south of the river did not. Since the heating was provided using coal, this was another natural experiment, allowing researchers to study the role of coal burning in deaths from respiratory diseases by comparing life expectancy for cities on both sides of the Huai River. This outcome was an unintended consequence of a policy that intended to provide inexpensive winter heat in order to provide basic comfort.[6]

China burns coal because it has coal, but it has just signed an enormous deal with Russia to import natural gas. Around the world, energy scholars have documented a phenomenon called the "energy ladder." As nations and individuals become wealthier, they seek out cleaner sources of energy;

these cleaner fuels are more expensive, but purchasers recognize that they cause less local pollution than burning lower-quality, less-expensive fuels. The energy ladder is an optimistic hypothesis that predicts that a wealthier China will seek out energy sources that create less local air pollution.

In May 2014 it was announced that China and Russia signed a US$400 billion, thirty-year deal that will pave the way for Russia to export a large amount of natural gas to China.[7] There were practical reasons for China to sign this deal with Russia: its government has plans to cut coal-burning power plants in a bid to tackle its huge air pollution problem. In April 2014 China's government set a target of 420 billion cubic meters of gas per year by 2020 to be part of its energy consumption mix. By importing natural gas, China will diversify its energy supply, shifting away from coal.[8] Those Chinese cities that switch from coal to natural gas for their winter heat and power generation will experience a reduction in ambient particulate matter, and the net effect will be a reduction in mortality risk and an improvement in quality of life.[9] This suggests that an unintended consequence of Russia's pivot from directing its exports of natural gas from western Europe to China could be a sharp improvement in urban quality of life in some Chinese cities. Recent evidence from Turkey supports this claim; there natural gas has been phased in as a major source of power generation, and the nation has sharply reduced its urban ambient particulate challenge and reduced its infant mortality rate.[10]

China is also seeking to diversify its import routes to prevent potential supply disruptions. In addition to the Russian supply, a huge pipeline project in northeast Myanmar will soon begin pumping twelve billion cubic meters of natural gas per year to the southern Yunnan Province. China also has three pipelines importing natural gas from central Asian countries, and in September 2014 president Xi Jinping visited Tajikistan to observe the start of work on construction of a fourth line, which will pass through Uzbekistan and Kyrgyzstan to deliver gas from Turkmenistan to China. When the new line is combined with the three other pipelines, China will eventually import eighty-five billion cubic meters of natural gas annually from central Asian countries.[11]

Given rising energy demand in the growing nation, increased environmental regulations, and the desire to increase the share of power generated by renewables, we cannot definitively predict China's future energy mix. We will present some predictions in the concluding chapter 10. The important point here is to note the fundamental trade-off: coal is inexpensive but dirty; natural gas is cleaner, but it must be imported from abroad. Renewable energy sources offer the promise of clean energy, but they currently are more expensive than conventional sources. In chapters 6 and 7, we will present evidence on the rising demand for less urban air pollution. This provides an incentive for government officials to consider public policies that shift the composition of power generation away from coal and toward cleaner sources.

Recent research by John A. Mathews and Hao Tan (2015) documents how China's electric power system is greening at the margins. Using data from 2014, on additions to China's electric power system, they note that China generated less energy from thermal sources in 2014 than in 2013 while increasing levels of water, wind, and solar power. Based on data on investment, Mathews and Hao reject the claim that China's power generation is getting blacker while its greening efforts remain small and insubstantial.[12]

The Evolving Geography of China's Industrial Production

Because of its mixture of past communist policies and the current strong hand of a centralized government and local mayors acting as entrepreneurs in guiding a booming economy, China offers a unique case for studying why and where industrial clusters emerge.

In the early days of the People's Republic of China, few countries (with the exception of the Soviet Union) supported the communist regime. At the time, the nation engaged in little international trade and its economic development was very slow. Due to national defense considerations, the central government's first priority was to develop inland industrial cities. In northeast China, Changchun, Harbin, and other cities preserved industrial facilities built during the Japanese invasion in the Second Sino-Japanese War. These cities offered a solid industrial foundation and provided support for military logistics, such as producing weapons and daily necessities after the Chinese Civil War. In the northwest, Taiyuan, Xi'an, and other cities were designated by the Soviet Union as key sponsored industrial cities. The central government invested large amounts of resources to develop these two regions, and ordered a large workforce to move there.

During the era of Mao Zedong, industrial specialization patterns were not determined by market forces but instead influenced by political considerations. In the late 1960s, for example, to prepare for possible wars with neighboring countries and regions there was a drive to relocate production of key industrial products from coastal areas to interior provinces. By moving industrial activity away from other nations' borders, Mao's policies represented a type of insurance policy against attack. In recent decades, South Korea has pursued the deindustrialization of Seoul as a national strategy to minimize the economic impact of an attack from North Korea.[13]

The key feature of China's industrial geography under Mao was that each region had a similar industrial composition and produced goods to achieve self-sufficiency. There were very few interregion flows of goods or people. During this planned-economy period, Mao promoted the concept of self-reliance. Unlike capitalist nations that embrace the concept that free trade and specialization raises overall income, China sought to minimize cross-provincial inequality. This was particularly beneficial to the poorest provinces. Such self-reliance boosted national security, as it reduced the economy's vulnerability to a shock in any one region.[14]

During the 1970s, as relations with the Soviet Union worsened, China came into closer contact with other countries, such as Japan, the United Kingdom, and the United States, and its development focus shifted to its coastal regions. Since the 1980s, thanks to economic reform and the strategy of opening up the economy in order to export to other nations, China has become the world's "factory," and its industrialization has largely been driven by the fast growth of export-oriented, labor-intensive industries in the coastal areas. In 2004, more than 90 percent of total exports and roughly 60 percent of industrial output came from coastal cities such as Shantou and Xiamen.[15] The old inland industrial cities in northern China declined.

Starting with the economic reforms in the 1980s, market forces have affected the spatial distribution of economic activity. Firms now cluster in places that offer the highest profits. An industry that exports to the United States, for example, will seek out a location close to a port with low industrial energy prices and access to inexpensive land and labor. As China's industrial production clustered close to coastal areas, and as the *hukou* system of internal passports has been relaxed, millions of industrial workers have migrated to eastern and coastal cities because of the job opportunities there.

But these coastal cities face growing costs. Over time, labor and land costs have risen. The quality-of-life costs in these cities has also risen as pollution emissions from industrial production have soared. A large number of industrial cities such as Beijing, Lanzhou, Taiyuan, and Wuxi suffered from severe air and water pollution in the early years of the twenty-first century.

Deindustrializing in High-Cost Coastal Areas

An emerging trend taking place across China is manufacturing's move away from the wealthy coastal megacities. Profit-maximizing factories compare their profits at different geographic locations and choose the location that offers the highest expected long-run profits. In Mao's era there were no land or real estate markets, and in the centrally planned economy, it was unlikely that scarce land was used for its most valuable purpose. Such land was inefficiently allocated so that manufacturing factories with low-density buildings and warehouses were located at the city center, and each factory allocated some of its land to build dormitories for workers. There are valuable alternative uses for such scarce land; it could be allocated through an auction system to the economic actor who values it the most. For example, the center city land that a Mao-era factory sits on could be redeveloped as a fifty-story residential or commercial tower.

This transition is now taking place in China's hot coastal real estate market. In cities such as Beijing and Shanghai, the price of land is very high, and manufacturing is land-intensive. High land prices motivate developers to seek out every opportunity to develop commercial and residential towers in the central city, and they are willing to pay high prices to the already existing manufacturing firms. Figure 2.4 highlights the growing land-price gap for industrial use in different regions after 2008.

RMB/m²

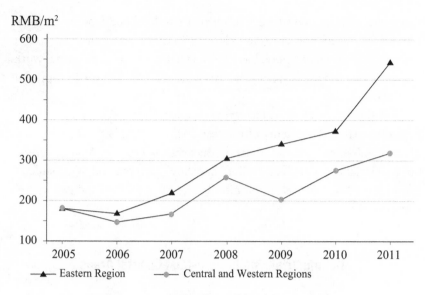

FIGURE 2.4 Regional industrial land-price trends

In China's major cities, former industrial sites are being rehabilitated and converted into skyscrapers. Manufacturing's share of total output in Beijing dropped from 30.7 percent in 2004 to 22.3 percent in 2013, while the service sector's share of employment increased from 67.8 percent to 76.9 percent. Similarly, manufacturing's share of output in Shanghai also dropped from 48.2 percent in 37.2 percent, while the service sector's share of employment increased from 50.8 percent to 62.2 percent during the same period. These manufacturing firms are now located in the nearby Hebei Province.

Local officials seek to have the best of both worlds; they want scarce prime land in the major cities to be allocated to real estate developers, but they also seek to keep the factories in their provinces in order to create low-skill jobs and collect tax revenue. The coastal provinces have shut down heavily polluting factories and relocated them to the relatively poorer regions within the same provinces.[16] For instance, Guangdong Province is subsidizing polluting firms in the Pearl River delta to relocate to the northern part of the province, and Jiangsu Province is relocating such firms to the north Jiangsu (Subei) area. This industrial relocation strategy allows provincial governments to green the big city while helping underperforming areas within the province to grow, but it also increases pollution inequality: air quality improves in the wealthier part of the province yet it may decline in the poorer parts. Whether this strategy of moving industry within a region actually improves big-city air quality is an important question. We recognize that regional emissions may blow back and degrade quality of life in the major city, but our own calculations suggest that this effect is quantitatively small: a 10 percent increase in neighboring cities' total industrial smoke emission is associated with just a 1.7 percent increase of local PM_{10} concentration.[17] If the spatial distance increases between the big city that is deindustrializing and the current locations of the displaced factories, this "downwind" effect shrinks further.

This relocation of firms vacated large parcels of center city land in Changsha, Chengdu, Dalian, Guangzhou, Shenyang, and Wuhan. In Beijing, many pieces of newly released land attracted real estate developers because of their excellent locations near Beijing's central business district. For instance, Beijing No.2 Chemical Plant vacated fifteen hectares of land there, which was worth more than four billion yuan based on a 2009 auction. Today, this land is almost completely filled with commercial and residential towers.

Another case is the relocation of polluting firms before the 2008 Beijing Olympics. The Chinese government promised to make Beijing an "ecological city" with "green space, clear water, and blue skies" after it won the 2008 Olympic bid. As one of the efforts to improve the city's air quality, in 2005 Beijing Capital Steel Group (Shougang) had begun to relocate its facilities to Hebei Province, some two hundred kilometers east of Beijing. This big steel producer reduced emissions of its sulfur dioxide, soot, and dust by 49.18 percent, 50.32 percent, and 49.22 percent, respectively, between 2007 and 2008.[18]

Over the last decade, the Beijing government relocated hundreds of firms out of the city center to reduce pollution. Those firms were willing to leave because they received considerable financial compensation that helped them overcome the financial difficulties they experienced during China's reform of its state-owned enterprises in the late 1990s and early years of the new century. Over the last decade, the Beijing municipal government relocated roughly eight hundred manufacturing state-owned enterprises within the fourth ring road to a remote suburban area and nearby Hebei Province.

In addition to high land prices, the megacities also have high wages. Around 2004, the average wage in coastal cities was five times that in inland cities. Since then, sharply rising wages in the coastal cities have pushed many labor-intensive manufacturing firms to move. Coastal cities also have a cost disadvantage with respect to electricity prices; for instance, in 2012 the industrial electricity price in Zhejiang Province was roughly 15 percent higher than that of Henan Province and 7 percent higher than that of Shaanxi Province.

At the same time that land prices and wages are rising, environmental regulation is also becoming more stringent in the coastal cities; this is not surprising, because these are China's wealthiest cities. The international environmental economics literature has documented a positive correlation between national per capita income and the intensity of environmental regulation: poor nations do not enforce environmental regulations, while middle-income nations and wealthy nations do. An explanation for this empirical regularity is that wealthier people's basic needs—such as access to food and shelter—have been satisfied and they now seek a higher standard of living and thus demand a higher quality of life; their government officials respond to this demand by introducing credible regulation. (We will return to this point in chapter 6.) In the case of China, stricter environmental regulation is making it more expensive for polluting industries to locate themselves in wealthy cities.

While the cost of industrial production in coastal cities is going up, the ability to produce in other locations has been improved. In recent years, China has made major investments in intercity roads and transportation networks. Today the length of China's national highway network stands at 104,468 kilometers, which exceeds that of any other nation. In 2013, roughly 90 percent of large- and medium-size cities were integrated into this highway network. Even a highway in Tibet (known as the Roof of the World), which will be linked to Beijing and many other cities, is under construction. As China invests in better transportation networks, it is likely that heavy industry will relocate to cities offering cost advantages.

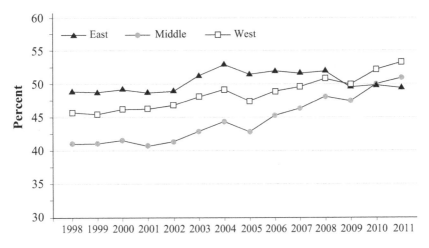

FIGURE 2.5 Industry's share of the overall economy by region

Without a good transportation network it is very costly (measured in time and money) to ship production outputs and consumer goods long distances. When transportation costs are high, manufacturing will tend to locate in big cities close to end consumers and ports. An environmental consequence of this choice is that millions of people in these cities are unintentionally exposed to the air and water pollution produced by these factories.

The net effect of trends regarding urban land and labor costs, government policies, infrastructure investment, and differential environmental regulatory enforcement is that dirty manufacturing is leaving the big coastal cities because it is becoming increasingly expensive to produce there.[19] In a 1999 paper, one of the present volume's coauthors, Matthew E. Kahn, documented that such deindustrialization is an important factor in explaining air quality improvement in US cities such as Pittsburgh.[20] The same logic holds in China. Our own calculation shows that a 10 percentage point reduction in the share of manufacturing jobs in a city will reduce its PM_{10} concentration by 3 percent.[21] Figure 2.5 documents how China's wealthier eastern region industry began playing a smaller role in the overall economy in 2004 and, beginning in 2005, industry's share of the overall economy grew sharply in the central and western regions, highlighting the regional divergence taking place across China's cities.

Does Manufacturing Migration Create a Zero-Sum Pollution Game?

As factories migrate from wealthy cities to poor cities, will a decrease in pollution in the wealthier cities simply reflect pollution increases in poorer cities? Similar to a game of poker, where what the winners win equals the sum of what the losers lose, is this a zero-sum game such that manufacturing migrates to poorer areas with weaker regulation? If this dynamic plays out, "a tale of two cities" could emerge, as the wealthier, more sophisticated cities become greener while a "browning" of poorer cities in China's western region takes place.

Consider the case in which a Mao-era dirty factory with outdated technology closes and moves from a highly populated city to a smaller one. In this case the aggregate public health costs would decline because fewer people would be exposed to the pollution. For example, if twelve million people live in Guangzhou and three million live in Shaoguan, a small inland city in northern Guangdong Province, public health is improved by moving a steel factory from Guangzhou to Shaoguan. Second, established factories do not literally stand up and walk to a different city; when an older, inefficient factory is closed in one city, a new factory is built in the destination city. Given ongoing technological advances, the new factories will be cleaner than the factories that are shut down—pollution and energy consumption per unit of output produced will be lower. This means that the migration of a dirty factory from a large city to a smaller city both reduces the net count of people exposed to the pollution and the per capita exposure in the smaller city, because the new factory is cleaner than the dirty factory that closed. The ongoing geographical dynamics of industrial production thus might not be a zero-sum pollution game. Instead, such production may generate net social benefit as high-density coastal cities deindustrialize (and hence reduce their population's exposure to pollution) while new factories in inland cities employ cleaner technology.

Our own calculation shows how a region's overall emissions from industry can decline even as economic growth is taking place.[22] In table 2.2 we give the results of a simple decomposition exercise on energy intensity (energy consumption per gross domestic product) from the industrial sector in eastern versus western regions. The table shows that the

TABLE 2.2 Energy consumption and energy intensity changes

	Year/period	Coastal cities	Inland cities
Manufacturing share in output	2006	46%	41%
	2011	44%	45%
Composition effect (TCE per million RMB)	2006–11	–3.17	6.83
Technique effect (TCE per million RMB)	2006–11	–22.4	–19.8

Note: When we calculate the composition effect, we assume sector-specific energy intensity is constant (using nationwide numbers in the initial year 2001) and let a city's industrial composition change over time. Similarly, when we calculate the technique effect, we assume a city's industrial composition is constant (using numbers in initial year 2001) and let nationwide sector-specific energy intensity change over time.

composition effect increases inland cities' energy intensity while reducing it in coastal cities. This means that eastern cities' economies are shifting away from polluting industries toward cleaner industries while the opposite is taking place in western cities. Between 2006 and 2011, the change in manufacturing share in output contributed to a 3.17 ton of coal equivalent (TCE) decline and a 6.83 TCE increase in coastal and inland cities, respectively. The good news is that technology upgrading made a major contribution to reducing energy intensity in both regions. The inland cities enjoyed a 19.8 TCE energy intensity decrease, which outweighed the positive composition effect so that the net change in energy intensity there is negative.

Over time, China's reliance on manufacturing's share of total output in gross domestic product (GDP) is likely to shrink for the nation as a whole due to rising wages, alternative uses for land, and increased environmental regulations. In 2012 manufacturing accounted for over 45 percent of China's GDP, and this figure is expected to drop by 10 percentage points by 2050.[23]

The environmental implications of such an industrial transition have been seen before in the case of the US Rust Belt, which, since the 1970s, has lost hundreds of thousands of manufacturing jobs as they moved to the US South and abroad. A "silver lining" of this trend is that cities such as Chicago and Pittsburgh have experienced urban river water quality and air quality improvement, and infant mortality has declined as a result.

These environmental gains were not a "free lunch," however; older, low-skilled workers who were employed in the now defunct manufacturing plants suffered a sharp cut in wages when reemployed elsewhere.[24]

As dirty factories close, the polluting emissions decline, but a factory's years of production can leave a scar on the immediate area surrounding it. In the United States, defunct chemical storage areas and industrial facilities are known as brownfields. A long-running debate has focused on who should pay to clean up these sites so that the land can be redeveloped for new uses. During China's rapid urbanization process, many old, dirty firms moved away from the central areas of large cities; thus, in recent years land contamination has emerged as a new and serious issue. The World Bank estimates that roughly one-fifth of vacated lands are heavily polluted. The total number of brownfields in China is more than 300,000.[25]

In 2001 George Akerlof was awarded the Nobel Prize in Economics for his work on the implications of asymmetric information on markets. In an ingenious paper, he focused on the market for "lemons" in the market for used cars. In this market, sellers (the current owners) have more information about cars' true quality than do potential buyers; buyers recognize that they "know that they do not know" the true quality of the cars. Akerlof demonstrated that there is an adverse selection challenge in that an owner of a used vehicle is most likely to sell it if it is a "lemon." If buyers anticipate this, the whole used vehicle market could unravel, because potential buyers do not want to overpay for a low-quality car, but the owners of high-quality cars have no way to signal the quality of their vehicles.[26]

This same "lemons" issue arises with respect to past industrial sites that are now being converted into high-rise residential and commercial buildings. Given that toxic waste remediation is a costly activity, local officials have an incentive to overstate the amount of remediation they invest in. Consider a case in which such officials engage in no remediation but convince the public that they have; the officials will save money (because they have not cleaned up the site), and they will collect more revenue when a land parcel is auctioned off because developers now believe it is a greenfield. Such an action would be more likely to take place if there is lag between exposure to toxic substances and future health impacts. In this case, the urban officials could argue after the fact that their actions did not cause any sickness that has arisen. This discussion raises the issue of accountability within a governance system: Who monitors the local officials to see if they are implementing the policies they claim they are? We will return to this topic when we discuss the Chinese central government's policies in chapter 8.

A Hong Kong Case Study of Reducing
Manufacturing Pollution's Local Impact

Manufacturing's environmental impact often does not respect political boundaries; air pollution can drift across regions, and industrial water pollution can flow downstream. In cities such as Beijing and Hong Kong, much of the air pollution is caused by emissions from nearby regions. This cross-boundary problem is most severe when a city is adjacent to a very dirty neighbor, such as Hebei Province near Beijing and Guangdong Province near Hong Kong. Roughly 30 percent of Beijing's particulate matter pollution (measured as $PM_{2.5}$) blows in from Hebei Province, where many polluting steel plants actively produce. But as the distance between the emissions source and destination cities increases, this cross-boundary effect attenuates quickly.

Hong Kong has suffered from nearby industrial emissions drifting into its airspace. In the Pearl River delta, about two-thirds of air pollutant emissions are from industrial cities in Guangdong Province such as Dongguan and Foshan. Hong Kong's air quality is also largely affected by regional industrial pollution imports, which lower Hong Kong's quality of life. The irony is that some dirty firms in Guangdong are in fact funded by capital investments from Hong Kong.

As Hong Kong's population and per capita income grows, the city's total willingness to pay to avoid industrial emissions increases. Hong Kong's population is roughly 7.3 million people. If the average resident is willing to pay fifty US dollars per year for an improvement in air quality, the people as a whole would be willing to pay $365 million per year to polluting factories to reduce their pollution. This is a contrived example to show how much money can be collected to tackle pollution when there are millions of prosperous people being affected by it. As the city's residents grow wealthier, this willingness to pay to avoid pollution would rise further. The simple economics of program evaluation states that a pollution mitigation project is good if its benefits exceed its costs. So, in this example, if there is an investment in pollution abatement equipment that costs less than $365 million per year, such a project would improve the aggregate well-being of the people of Hong Kong.

In fact, Hong Kong's government has already begun to implement this approach; since 2008 it has paid manufacturers in Guangdong Province about $130 million every year to install pollution-reducing equipment. In 2008 the Hong Kong government launched the five-year, $93 million Cleaner Production Partnership Program in collaboration with the Guangdong government. This program subsidizes Hong Kong–owned factories that locate in Guangdong to install pollution-reducing equipment and adopt cleaner production practices. By the year 2010, this program had already yielded significant environmental benefits, as indicated by the reduction air pollution relative to 1997 levels; significant regional reductions (of 20 to 55 percent) have been achieved for sulfur dioxide, respirable suspended particles, mono-nitrogen oxides, and volatile organic compounds.[27] The Hong Kong–Guangdong Joint Working Group on Sustainable Development and Environmental Protection endorsed a new regional air pollutant emission reduction plan in November 2012, setting out specific targets for 2015 and target ranges for 2020; the two sides will review progress in 2015 to finalize the targets for 2020. Following Hong Kong's successful experience in reducing pollution, the Guangdong government also invested in local firms to reduce their emissions. This policy will continue until at least 2015, and the Guangdong government is also moving many dirty firms away from Hong Kong.

On mainland China, Beijing has recently paid compensation to factories located in nearby Hebei Province to shut down or relocate to other provinces. Environmental scientists at the Chinese Academy of Sciences estimate that roughly 30 percent of Beijing's pollution problems are caused by imported industrial pollution.

Cross-boundary pollution challenges arise within the region as China's coal plant emissions of sulfur, mercury, ozone, and particulate matter move downwind across the East China Sea, affecting Japan and South Korea. In November 2013, the South Korean media began to refer to the smog making its way from China as "air raids." A Japanese study claimed that air pollution from China was responsible for the high level of mercury deposited on iconic Mount Fuji.[28] Some cooperation in addressing air pollution is already taking place; for instance, Japan has given the Shanxi provincial government a loan of US$125 million to subsidize the desulfurization of large coal-fired plants in the capital city of Taiyuan. In December 2014, representatives of the three governments held a two-day summit just outside Beijing and vowed to work cooperatively to combat the region's polluted air.[29]

These examples from Beijing, Hong Kong, Japan, and South Korea highlight how bargaining with well-established property rights can lead to improvements in overall economic efficiency. In 1991 Ronald Coase won the Nobel Prize in Economics. The Coase theorem posits that when property rights are well defined and bargaining costs are low, the unintended victim of someone else's actions (those exposed to pollution, for example) can negotiate a mutually beneficial transaction with those causing the problem (the polluters) as a means of mitigating the problem.

Three Other Environmental Challenges
Posed by Industrialization

We have thus far focused on air pollution and greenhouse gas emissions associated with manufacturing. In this section we discuss three other environmental challenges affected by industrial growth. Our specific interest is in the spatial aspects of these challenges and in understanding their root causes. Such an understanding is a necessary step for discussing the likelihood of political solutions to be adopted (which we discuss in chapters 8 and 9).

Industrial Water Pollution

The city of Kunming, in southwest China, is considered one of the most pleasant cities in the nation, with springlike weather all year round. In recent years, however, industrial water pollution has become very serious. Many mining firms along the Xiaojiang River discharged wastewater directly into the river and made it noxious, and the local media have dubbed it

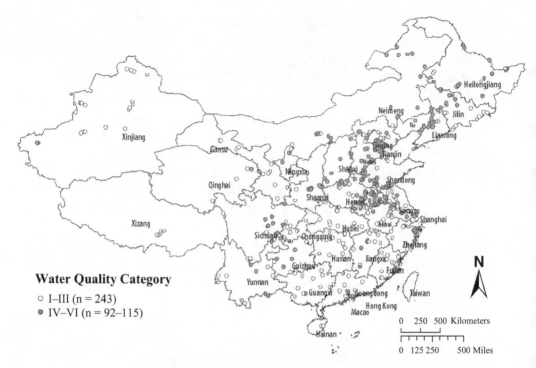

FIGURE 2.6 The spatial distribution of water quality
Source: World Bank, *Cost of Pollution in China* (Washington, DC: World Bank, 2007).

the Milk River because the contaminated water has turned white. Due to the lack of clean water, farmers have had to use this milky white water to irrigate their fields, and this destroyed many of their crops. Another terrible water pollution case happened in Dian Lake in Kunming, which has experienced cyanobacteria bloom problems since the 1990s; after twenty years this problem has still not been solved.

In 2004, about 58 percent of the seven main rivers and 74 percent of twenty-seven major lakes in China contained water graded class IV, V, or worse and were deemed unsafe for human consumption.[30] Figure 2.6 shows the spatial distribution of water quality levels in China. The dirtiest sections of the seven rivers are mostly concentrated in areas with manufacturing and mining activities, such as Hebei Province (near Beijing), Inner Mongolia, and Zhejiang Province.

Like many forms of pollution, water pollution does not respect urban or international borders. Industrial emissions from a certain location can flow downstream and become someone else's problem, and this reduces the incentive for urban leaders to regulate their jurisdictions' own profit-generating firms. Environmental economists have documented this "free rider" problem along major European rivers. In western Europe, pollution along the Danube River is higher at international monitoring stations located where two nations meet rather than in the interior of a nation. Some nations are locating dirty industry near borders so that they gain the advantage of economic activity, but the environmental costs, whether from industrial or wastewater treatment, are borne by neighboring countries downstream.[31]

Research based in China has also examined a similar free rider problem, documenting cross-boundary river pollution effects caused by seven primary Chinese industries: agricultural products and by-products, textile manufacture, garment manufacture, pulp and paper, petroleum and nuclear fuel processing, chemical manufacture, and nonferrous metal smelting and pressing.[32] Such cross-boundary pollution poses key quality-of-life issues because it suggests that local mayors have no direct incentive to address the pollution challenge because they collect tax revenue and workers in their jurisdiction enjoy better employment prospects; in the meantime, however, downstream neighbors bear the environmental costs.

Industrial Growth Polluting the Domestic Food Supply

New pollution challenges are emerging in rural areas located close to new industrial facilities. The *Wall Street Journal* has reported that industrialization in the Chinese countryside is threatening the food supply.[33] In an article focusing on Dapu, a rural area in the middle of China's breadbasket, the Hunan Province, a farmer complains that a state-backed chemical factory next to her farm dumps wastewater directly into the local irrigation pond. One farmer there reports that the rice she grows cannot be sold because of its low quality, but she grows it anyway to qualify for compensation payments from factory owners. Yet this compensation, the farmers claim, does not meet their lost revenue caused by the pollution damage.[34] Forty-four percent of rice samples were found to contain poisonous levels of cadmium in Guangzhou, the capital city of Guangdong Province in southern China, and that rice was being served in various restaurants around Guangzhou. The cadmium-laced rice is a reflection of the heavy levels of heavy metal pollution that can be found throughout China's farm lands. It is estimated that the country loses US$3 billion per year to soil pollution, and that between 40 percent and 70 percent of China's soil is already contaminated with heavy metals and toxic fertilizers.[35]

This case highlights the point that ecologists often stress: that there is a "web of nature" connecting the urban economy to its natural surroundings. If polluting factories leave the city and relocate in the countryside, urbanites enjoy a reduced direct exposure to industrial pollution, but they increase the risk of contamination from domestic food as dirty industry moves near farmland.

One market solution to this problem would be to construct a standardized and detailed labeling system that would signal to customers where the food is grown and what standard it reaches. If urban consumers are aware that dirty factories are located in an agricultural area in Hunan Province, they may choose not to buy rice from that area. Still, poor people may be more likely to ignore these labels, as they may not have the money to be able to afford more expensive, safer rice. But the threat of losing the market share from sophisticated consumers will lead food growers to increase their physical distance from dirty rural factories. Such a "moat effect" would increase the quality of the food supply. We provide this example to highlight how the growth of the urban middle class can encourage firms to differentiate their products as they compete for market share; higher-quality products can fetch higher prices. If food pollution risk is one determinant of quality, those farmers who are producing high quality products will have an incentive to differentiate their products. At US supermarkets such as Whole Foods Market, organic produce sells for a price premium relative to conventional produce; this same situation could play out in China. This example is meant to highlight how free-market competition can actually cause an increase in sustainable farming when consumers have the information about the "greenness" of different products and when they are willing to pay a price premium for "green" varieties.

The Mining Cities

Rare earths mining and processing is another well-known example of a profitable industrial activity with significant environmental consequences. Ironically, rare earths are key components in green products, such as hybrid vehicles, wind turbines, and solar panels. China is the world's leading exporter of these metals, with a world market share of 95 percent, and extracting them causes major environmental problems in the areas where they can be found, such as Inner Mongolia and Sichuan. Baotou, a city in Inner Mongolia, is known as the rare-earth capital of the world; residents there inhale solvent vapor, particularly sulfuric acid, as well as coal dust, which is clearly visible in the air. The soil and groundwater are full of toxic substances, and this sharply reduces farming output.

The miners exposed to this pollution suffer sickness, and there is increased mortality risk; many workers in the mines die from cancer in their thirties, possibly from exposure to radioactive materials. Every day they breathe in the dust, and they commonly suffer from pneumoconiosis, better known as black lung. As a result of these health risks, people have been migrating away from Baotou.[36] The miners are aware of the consequences of working in such conditions, but many still choose to work there because they can earn what they regard as higher "combat pay." Meanwhile, the mine owners get rich; they send their children to China's large cities or abroad for a better education and a higher quality of life, but they stay in the mining areas to earn more money.

Emerging Trends in Reducing Industrial Pollution

In this section we highlight several other promising trends that raise the possibility that the pollution associated with Chinese industrialization will shrink.

Formal Business Training and the Rise of Professional Managers

The quality of China's universities is rising, and the MBA graduates of such universities are receiving an education that will create a new generation of professional corporate managers. Economists have argued that better corporate managers have beneficial consequences for energy efficiency and, ultimately, environmental performance. Consider a profit-maximizing firm such as a steel factory that does not directly gain any extra profits from protecting the environment. If this firm has better managers in charge, these men and women may be better able to identify inefficiency in the organization and stay on top of technological trends (such as engineering breakthroughs) to match capital investments with their factories' needs. The net effect of such efficiency is to require less energy per unit of output. This offers private benefits to the firm because it needs to purchase less energy, and it offers social benefits because fewer fossil fuels will be needed to generate the same level of industrial output.

This optimistic hypothesis is supported in the academic economics literature. In a study of British industrial plants, a team of economists surveyed managers and created an index of "manager quality." In a second stage of the analysis they documented that, all else equal, firms with better managers had greater energy efficiency.[37] One possible explanation for this intriguing fact is that MBA-trained executives are skilled in data analysis, accounting, and finance, and these hard skills help them to identify which investments in energy efficiency are profitable. In this sense the computerization and benchmarking of industrial plants offers the possibility of improved environmental performance. In addition, these MBA-trained executives also have more specific knowledge on energy and the environment; for instance, in the MBA and executive MBA programs run by Tsinghua University's School of Economics and Management, students are taught such courses as Energy and Resource Management, The Trend of New Energy, and Energy and the Environment. Such specialized courses are more likely to be offered if there is a demand for increasing energy efficiency. For example, it can take more than a decade to train as a brain surgeon, and a young person will make this sacrifice of time and effort only if there will be a demand for such specialized skills. This explains why such individuals locate in big cities, because in such locations there is enough expected demand to justify this specific investment in human capital. This logic suggests that if more of China's factories seek to be energy efficient, more young people will invest their time in gaining human capital and skills that foster energy efficiency. The net effect of this trend will be a "greening" of aggregate industrial energy efficiency in China.

This managerial emphasis on energy efficiency is especially likely if executives believe that energy prices will rise over time in China. As energy prices rise, identifying strategies for economizing on energy consumption become more profitable. This matters for China because a common attribute of communist nations is deep energy subsidies. Low prices for key inputs help such nations industrialize, but they create weak incentives for their factories to be energy efficient. If China reduces its reliance on dirty but inexpensive coal and instead substitutes cleaner but more expensive natural gas and renewables, it is likely that industrial consumers and commercial buildings will face higher prices for electricity in China. This differential in the cost of generating power using fossil fuel versus renewables will continue to change over time as wind and solar become more cost competitive. Despite ongoing technological progress in designing efficient wind turbines and solar panels, it is likely to be the case that power generated from fossil fuels will continue to be less expensive. This means that any policy-driven substitution toward renewable sources of power generation will cause higher electricity prices. Such a rise in prices would incentivize firms to be more energy efficient.

Foreign Direct Investment and Foreign Knowledge Transfer

In our past research we used data from thirty-five major Chinese cities over the years 2003–6 and found that cities that receive more foreign direct investment (FDI) have less ambient air pollution.[38] A possible explanation is that FDI inflows provide Chinese factories with the capital, and perhaps complementary knowledge, to adopt wealthier nations' production techniques. The net effect is a reduction in the Chinese industrial sector's pollution per unit of production.

Our finding on the negative correlation between FDI inflows and pollution stands in contrast to a previous findings of a positive correlation, which is based on data from the 1980s and 1990s. Back then, FDI was used to build new factories, and this increase in the scale of industrial activity raised local pollution levels. Our 2010 study highlights that both the quantity of industrial activity *and* its quality are crucial in predicting likely pollution impacts.

Green Technology

Another promising trend in the local and global impacts of China's power production is China's ambition to be a major player in the nascent green global export market. China's export market share of solar photovoltaics increased from 2 percent in 2000 to 54 percent in 2011.[39] China and India are emerging as key providers of inexpensive equipment to the West for its production and consumption of clean, for renewable sources of energy. Technological advance combined with vast economies of scale in production open up the possibility that clean energy will become cost-effective. Such a composition shift could significantly reduce national greenhouse gas emissions associated with power generation. The expectation that China and India will play increasingly key roles in the supply chain for the mass production of green innovations is leading US venture capitalists to invest more in renewable power. Companies such as Walmart are committing to generating more of their electricity from renewable sources.[40] The expectation that such major corporations demand more solar panels and wind turbines creates a "field of dreams" effect such that exporting nations like China are aware that if they build high quality and inexpensive renewable technology, there is a demand for such equipment in high volume.[41] Competition among the world's suppliers of such renewable technology lowers prices and raises the likelihood that the total quantity sold will increase.

Greater international trade and Asian economies' increasing production of equipment for renewable power has played a significant role in reducing the cost of such equipment. For example, the average solar photovoltaic module price has fallen from US$4.66 per watt in 2004 to $2.01 per watt in 2010, and is expected to drop further, to $1.49 per watt, by 2015.[42] As China emerges as a leader in the nascent wind- and solar-power equipment markets, declining prices will provide the nation with an incentive to build fewer coal power plants and to increase its own reliance on renewable power.

Underlying China's surge in green technology is the nation's enormous investment in the human capital of its young people. In a recent National Bureau of Economic Research working paper, Richard B. Freeman and Wei Huang (2015) write,

> In the past two decades China leaped from bit player in global science and engineering (S&E) to become the world's largest source of S&E graduates and the second largest spender on R&D and second largest

producer of scientific papers. As a latecomer to modern science and engineering, China trailed the US and other advanced countries in the quality of its universities and research but was improving both through the mid-2010s. This paper presents evidence that China's leap benefited greatly from the country's positive response to global opportunities to educate many of its best and brightest overseas and from the deep educational and research links it developed with the US. The findings suggest that global mobility of people and ideas allowed China to reach the scientific and technological frontier much faster and more efficiently.[43]

Growth economists emphasize that ideas are public goods. An effective idea in one nation can be imported into China, and China's millions of trained college graduates help accelerate this diffusion process. The net effect of this technological imitation and experimentation is the possibility that globalization of ideas and trade can accelerate the sustainable power agenda. This optimistic logic stands in contrast to the usual claims made by many environmentalists that international trade damages the environment through pollution haven effects.

An open question concerns whether the trends we have discussed in this section add up to having a large impact on China's overall reducing of its greenhouse gas emissions per dollar of output. We will revisit this question in chapter 10, where we discuss the challenge for China of reducing its greenhouse gas impact during a time of rapid urbanization.

In this chapter we have discussed how the geography of Chinese industrial activity affects the spatial distribution of pollution. One major trend is that industrial jobs have attracted hundreds of millions of Chinese people to cities. Such industrial jobs offer a higher wage than people can earn in the countryside. In chapter 3 we will examine the trade-offs people face in making decisions about where to live, and we will show how this is likely to play a great role in China's environmental future.

CHAPTER THREE

The Migration to Cities

The Chinese central government recognizes that urbanization is a key for continuing national growth and to achieve improvements in the standard of living of its rural poor. The *New York Times* reports, "China plans to spend 40 trillion yuan (6.5 trillion US dollars) to bring 400 million people to cities over the next decade as the government tries to turn the country into a wealthy world power with economic growth generated by an affluent consumer class. Urbanization is a policy priority for China's government which wants to create a true consumer class that will help rebalance growth drivers away from the investment-heavy, export-oriented model it has followed for three decades."[1]

In this chapter we study both rural-to-urban and city-to-city migration. We focus on the city choice of the wealthy, the middle class, and the poor in modern China; these three groups differ with respect to their education and their aspirations. A successful nation would offer a high urban quality of life to each of these groups. This chapter explores the opportunities and challenges for each group; while chapter 2 focused on how China's industrialization affects the urban environment, here we seek to explain the trade-offs that Chinese urbanites face in choosing which city to live in; local environmental quality is one attribute among other priorities. For example, consider the choice of attributes that a migrant to Beijing is offered. Despite its high pollution level, Beijing continues to be a highly attractive destination for urbanites. For many individuals, including the present volume's coauthor Siqi Zheng, the opportunities provided by a unique capital city and its great universities compensate for the quality-of-life challenges of congestion and pollution.

One Urbanite's Story

Siqi hired a nanny named Ms. Fu to take care of her son. Ms. Fu was born in a small village in Henan Province. After she finished her junior high school education, she started working in the town near her village. She met her husband and got married when she was twenty-one and they had two sons. (In rural places one can violate the nation's one-child policy and pay a small fine.) The couple wanted to earn more money, so they decided to move to Beijing when she was twenty-five. Their two sons have been living with their grandparents in their home village; one son is now thirteen, and the other is eight.

Ms. Fu has been working in Beijing as a nanny for several families, and her husband works on construction sites. They rent a very small place (around eight square meters) in an urban village (an informal housing settlement) near Siqi's home. They are very frugal, and send as much of their earnings as they can back home to pay for their two sons' education and to support the rest of their family. Ms. Fu told us that, in their small village in Henan Province, almost all young and middle-age adults have left for big cities as "floating workers" for better earnings; only older people and youth remain in the village. During each year's spring festival, those floating workers return to their village for a half-month vacation.

Ms. Fu likes the vibrant urban life, and she is optimistic about the future, but she also complains that without access to a Beijing *hukou* internal passport she is denied access to basic public services for her children, such as health care, a pension, and educational opportunities; that is why her two sons are still back in their home village. Ms. Fu and her husband do not want to go back to that small village. After ten years' hard work, they saved some money and bought a new condominium unit in a third-tier city in Henan Province, and thus obtained that city's *hukou*, and their two sons were able to enter primary school and high school in that city.

Roughly sixty-one million Chinese children—one of every five in the world's most populous nation—haven't seen one or both parents for at least three months, according to the All-China Women's Federation, a Communist Party advocacy group.[2] Stanford University economist Scott Rozelle has documented high rates of mental health problems for rural Chinese children. One possible explanation for this is the long absences of their urban parents, who have little involvement in their children's daily lives. Families have been divided because many of the informal jobs are in Beijing and Shanghai, but the informal residents who lack internal passports do not have the right to send their children to schools in these cities, where local leaders have been slow to extend health and education benefits to the children of floating workers because it would be costly. Research from Brazil has argued that urban leaders intentionally choose to deny local benefits to poor migrants because they anticipate that if they offer such benefits this will only accelerate the migration of rural poor people to their cities.[3]

There are tens of millions of people like Ms. Fu across China who are moving to cities to earn a better living. By the year 2050 there could be a billion urban people in China, and the nation's urbanization rate could be about 70 percent. This means that, relative to today, there could be hundreds of millions more people living in China's cities.

China's Urban Growth Today

In 1978, China only had 193 cities, and the urbanites accounted for 17.9 percent of the total population. Today China has 657 cities, and 52.6 percent of Chinese people live in these cities; it is expected that 70 to 80 percent will live in cities by the year 2030.[4] The 2010 census reveals that 78 percent of migrants originated from rural areas. The three biggest gainers in terms of population between 2005 and 2010 as a percentage of their initial 2005 population were Beijing (the national capital), Shanghai (China's emerging commercial center), and Zhejiang (a leading manufacturing city), all of which are located in the east.

A widely used city classification is to group prefecture-level and other large cities into a first tier, second tier, and third tier. There are four first-tier cities—Beijing, Guangzhou, Shanghai, and Shenzhen—and they account for roughly 10 percent of all urbanites in China. The second-tier cities include all provincial capital cities (except Lhasa) and some major coastal cities with large populations; these cities account for roughly 30 percent of the total urban population. The third-tier cities include all other cities, though sometimes very small cities are further classified as fourth-tier cities. Between 1985 and 2010, the average annual population growth rates for these three groups of cities were 5.18 percent, 3.33 percent, and 3.60 percent, respectively. It is relevant to note that this growth was not fueled by urban residents having more children. Economists argue that urbanization has a causal effect in reducing fertility because women have more labor market opportunities in cities and urban housing is expensive; both factors encourage households to have fewer children. China's demographers predict that the nation will have a total population of 1.39 billion by the end of 2015.

Growth of Energy and Water
Demand Caused by Urban Growth

As a city's population grows, there is an increased demand for electricity and water. In chapter 2 we discussed at length the local air pollution and greenhouse gas emission implications caused by electricity generation. Due to spatial differences in rainfall, different geographic areas face different water supplies. Chairman Mao Zedong pointed this out during an inspection tour in the early 1950s, noting, "The south has plenty of water, but the north is dry. If we could borrow some, that would be good." The greater Beijing area faces the challenge of rising demand for urban water while not experiencing much rainfall, and this leads to the depletion of local aquifers. Large-scale water transfers have long been advocated by Chinese planners as a solution to the country's water woes. The South-North Water Diversion Project is a US$96 billion, 2,400-kilometer network of canals and tunnels designed to divert 44.8 billion cubic meters of water annually from China's south to its parched, industrialized north. In December 2014 the project's "middle line" officially began carrying water from the Danjiangkou Reservoir in central China's Hubei Province to Beijing—a distance equivalent of traveling from Corsica to London. Moving water is expensive and energy-intensive, and a variety of technological fixes such as water desalination are also being invested in. Ecologists have argued that an alternative strategy is to invest in infrastructure that captures rainwater and recharges aquifers and to engage in wastewater recycling.[5]

Southern California's recent experience offers relevant lessons. Tens of millions of people live in metropolitan Los Angeles and San Diego, a desert area where it rarely rains. To supply water to this vast region, huge investments have been made in canals and water delivery infrastructure. Despite the fact that it rarely rains in Southern California, many homeowners have swimming pools, and Los Angeles residents pay roughly a half cent per gallon of water. Economists have pointed out that in the United States people are rarely aware of how much water they are using during their daily shower or when they run their dishwasher. Users receive water bills only once a month, and many urbanites have signed up for automatic payment; they thus do not even see their water bills, as the expenditure is simply paid by credit card.

In the midst of the big data revolution, there is the potential to provide much more real-time information about household consumption. Higher prices would encourage direct conservation and in the middle term would encourage the purchase of water-conserving devices such as more efficient dishwashers and low-flush toilets. Higher water prices would lead Southern Californians to rip out water-guzzling lawns and plant more drought resistant plants.

The same logic holds in China's cities; water conservation through a price mechanism is the logical way to balance aggregate demand with local supply. Urban planners could engage in simple arithmetic, multiplying the number of people who live in a city by the typical consumption per capita and then extrapolating the city's future water demand using projections of aggregate population growth. If water prices are rising, this approach would overestimate water demand because such an extrapolation exercise ignores the law of demand at work. People consume less of a resource as its price increases.

The communist Beijing government is learning to use the price mechanism to encourage more efficient water use. On May 1, 2014, the Beijing Municipal Commission of Development and Reform raised the price of household water and started a new multitiered water pricing plan. A public meeting to discuss the changes was held before the announcement of this new water pricing policy. According to the plan, the lowest-tier water price rises from four yuan to five per cubic meter for households with an annual consumption lower than 180 cubic meters, which covers 90 percent of households. Households with an annual water consumption ranging between 180 and 260 cubic meters will be charged seven yuan (roughly one dollar) per cubic meter, and the water price for annual consumption over 260 cubic meters will rise to nine yuan per cubic meter. The cost of nonresidential water consumption will be adjusted accordingly. Siqi's three-person family is in the lowest tier, having only consumed fifty-four cubic meters of water in the first half of 2015.

While China's cities face many environmental challenges caused by rapid growth, it is notable that they have avoided the challenge of infectious disease commonly associated with dirty water. In the United States in the late nineteenth and early twentieth centuries, and in many developing countries today, there are cholera and typhoid epidemics associated with polluted water and the insufficient treatment and collection of sewage in growing cities. An advantage of a centrally planned economy is the ability to mobilize everyone and make these expensive top-down investments. Beginning early in the 1950s, the Chinese Communist Party began to mobilize the population to engage in mass "patriotic health campaigns" aimed at improving the low levels of environmental sanitation and hygiene and at attacking certain diseases. One of the best examples of this approach was the mass assaults on the "four pests"—rats, sparrows, flies, and mosquitoes—and on *Schistosoma*-carrying snails. Particular efforts were devoted in the health campaigns to improving water quality through such measures as deep-well construction and human waste treatment.

Air-Conditioning in Chinese Cities

As China's urbanites grow wealthier, they seek out the basic comforts that people in the United States take for granted. China's summers are hot and humid; air conditioner use is skyrocketing in the cities. Now 44 percent of Chinese urban households have air conditioners, while this share is only 12 percent in the rural sector. In a recent paper, Lucas Davis and Paul Gertler studied air-conditioning demand in Mexico and used the results to extrapolate what could be the increase in global world demand for electricity in a warming world with wealthier urban households.[6] They document that wealthier people and those who live in hotter climates use more air-conditioning.

When Siqi was a teenager, her home did not have an air conditioner. Her family used fans during the hot summer, and sometimes they slept on the floor with a summer sleeping mat during very hot days. Her parents bought their first air conditioner in 1998. Today in Siqi's community all households have air conditioners, and Siqi has three air conditioners in her current home. All the classrooms and student dormitories at Tsinghua University also have air conditioners.

Housing Market Dynamics in China's Cities

An individual's decision on which city to move to depends on local home prices. In recent years housing price appreciation in several Chinese cities has been phenomenal. Beijing has had an annual average appreciation rate of 27.4 percent, whereas the average annual appreciation rate is 14.3 percent for thirty-five major cities during 2006 and 2013.[7] Tsinghua graduate students and their spouses have to save for thirty years in order to buy a hundred-square-meter condominium unit in Beijing; but if they choose to work in Xining, this couple could purchase a unit after saving for ten years, or possibly even less time.

The most stressful challenge for young Chinese urbanites is soaring real estate prices, especially in large cities. Renters faced with such staggering price increases have limited purchasing power even if they come from the upper middle class. Siqi is lucky; she bought her first condominium in 2003 when she was still a doctoral student at Tsinghua University. At that time the price was five thousand yuan (roughly US$700) per square meter. In 2013 real estate in the condominium tower where she lives was priced at roughly forty thousand yuan (roughly US$6500) per square meter. At a time when most US residents are earning less than 1 percent interest on their savings accounts, the annual average rate of real estate appreciation in Beijing has been 30 percent. This means that home prices more than double every three years. Strong housing demand from fast urbanization, constrained residential land supply manipulated by local governments, and the shortage of alternative investment opportunities are the main explanations for this "golden period" in China's real estate over the past ten years. Those who did not buy condominiums before this boom regret it, as their income during the ensuing years has not kept pace with the growth in housing prices. Today Siqi's students face a high cost of living Beijing; two of them bought their new condominiums with significant financial help from their parents; three others rent small apartments.

Real estate economists use the supply-and-demand model to explain housing prices across cities at a given point in time and to explain pricing dynamics over time. A standard analysis of the factors that influence a city's aggregate housing demand include population growth, income growth, interest rates, and home buyers' expectations of future price appreciation.

Some additional demand factors apply in China's urban housing market. Chinese households have fewer alternative investment opportunities because they have limited access for investment in foreign assets. The domestic stock market is viewed as a very risky investment; a common belief is that only those who can access inside information are able to make money there and most small investors are losing money. That is why a huge amount of money flows into the housing market—after all, everyone needs a house to live in!

A more novel reason for why Chinese house prices have been rising focuses on China's skewed gender ratios. There are roughly 103 men ages twenty to thirty-four for every 100 women of that same age in Chinese cities. Young men pursue home buying because it raises their prospects of marrying. The combination of the nation's one-child policy and the high male-to-female ratio in big cities provide strong social incentives for young men and their parents to save up money to pay the down payment for condominiums in order to raise the likelihood that the young men can get married.[8] If parents and their children believe that housing prices will keep rising, these young people will seek to buy a home at an even earlier age.

Those male migrants to a new city whose parents did not leave a house for them face great pressure to earn and to save. The Chinese media has stressed that young men seek to buy cars and condominiums to impress young Chinese women. Given that the real estate prices in Chinese cities have been soaring in recent years but car prices have been stable, a house means more for young Chinese women and their parents. In this sense, housing is an input in the production of a young man's status, and increased status raises his marriage prospects. In modern China, young women are in scarce supply and in the absence of immigration their numbers will not increase for decades. In the case of finding a spouse, young men face a zero-sum game competition, and this adds to the stress of urban life.

China's Superstar Cities

China's new urban upper-middle-class households do not tend to be in the industrial cities. Though only about 15 percent of Tsinghua University students originate from Beijing and Shanghai, every year more than 60 percent of graduates go to work in these two superstar cities. This pattern has persisted over the last fifteen years.

The clustering of the skilled in specific cities and specific locations within those cities drive up land prices, and this means that the poor cannot afford comfortable housing in such areas; this induces residential segregation. An offsetting phenomenon is that the wealthy need to hire the poor to live close to them in order to provide such services as preparing food and cleaning their clothes. Through trading these services the poor are selling their time to the wealthy. This is why we can still observe that poor people live and work in those superstar cities, though most of them reside in low-quality but inexpensive housing communities. Similar to what has been observed in recent years in New York City, rising real estate demand and limited supply means that home prices rise and the middle class is squeezed out.

In this sense, China's superstar cities' income dynamics mirror the trends playing out in the United States. In the recent past, most college students in Beijing and Shanghai used to remain in those cities after graduation, but today, unless you hold a master's or doctorate from an elite university in one of these big cities you are unlikely to be earning a high enough wage to be able to afford a condominium there. Extremely high housing prices in Beijing and Shanghai lead to a situation where these cities are populated with high-income people and the low-skilled people who work for them; this contributes to high income inequality and spatial segregation.

A negative consequence of the success of China's superstar cities is that their standard of living has diverged from the rest of the nation. Given China's communist tradition, this sharp increase in inequality has caused both social dislocation and tension. One of Siqi's graduate students witnessed this inequality with respect to the quality of his high school education and how this affected his likelihood of being admitted to China's best universities. Cong was fortunate that his parents brought him to Beijing in 2002 to study at a famous high school. He sensed that he received a much more enriched education at this school relative to his childhood friends who remained in his small hometown. In Beijing he and his new friends learned to swim and conducted interesting biochemical experiments every week. He took seven classes per day and enjoyed free time during the weekend. His old friends, from the small town, had to take more than ten classes per day and only had a half day of free time on the weekend.

Cong entered Tsinghua University three years later, but his childhood friends, despite working extremely hard, were less likely to be admitted to the elite colleges. Under the college entrance quota system, big-city high school students have an advantage over those from other areas in being admitted to the elite schools. In 2013 China's prestigious Tsinghua University enrolled about fifty-one students out of every ten thousand Beijing high school students who took the college entrance exams, whereas this figure drops to 3.6 and 2.4, respectively, for students from Henan and Sichuan provinces.[9] This admissions gap persists even if we adjust the statistics to account for the fact that the elites tend to live in the superstar cities and thus their children are disproportionately reared in these cities. Such inequality in educational opportunities persists, and this has greatly angered parents who live outside of the major cities.

China's central government understands the concern about this inequality in educational access and it has been taking steps to narrow the regional gap in college enrollment to ensure that students from different regions have more equal access to higher education. Many new colleges have been built in the western and central regions, but their reputations cannot be created overnight, and it is difficult for those new colleges to attract excellent professors.

Shanghai

Though Shanghai is the commercial and financial center of mainland China, it still exudes charm—notably with the Huangpu River, which runs through the city center, and European-style buildings from the days of American and European settlements. The entire world has discovered Shanghai's new face and its new construction thanks to the 2010 World Expo and the iconic nighttime scene showcasing its skyscrapers in the James Bond movie *Skyfall*. It is the most Western-influenced Chinese city.

Shanghai is a bustling city of about 17.7 million people, and living there means life in the fast lane. Everything is hectic: the people, the traffic, and the bicycle traffic. As China's industrial and financial center, Shanghai is struggling with problems typical of life in other Chinese megacities, especially Beijing: environmental problems like smog, water pollution, and noise, as well as overpopulation, traffic jams, and a housing shortage. Despite these urban problems, Shanghai remains a glitzy "Paris of the East." Among the top ten most expensive cities in the world for expatriates, Shanghai is a consumerist mecca. Its cosmopolitan atmosphere includes vibrant nightlife and a major arts and entertainment scene.

Shanghai's local government recognizes that taking care of citizens' quality of life will attract larger amounts of capital and a greater number of high-skilled people. In early 2013, the whole nation was astonished when sixteen thousand dead pigs came floating down the Huangpu River. It was a classic tragedy of the commons: improper pig disposal upstream, at Shaoxing, was now causing pollution of Shanghai's river.[10] Later that same year, Shanghai encountered serious haze due to the exhaust emissions from coal-fired power plants. Since then, environmental issues in the city have received great attention; Shanghai's mayor has announced his intent to combat pollution in an effort to create a better city and give its residents an improved quality of life.[11]

Shanghai is home to a number of industrial parks that recruit major firms to colocate in close physical proximity. Such agglomeration fosters increased productivity as plants located close to each other learn and trade with each other. A major technological cluster has emerged in Shanghai's Pudong District, which was formerly a marshland where Shanghai's vegetables were grown. Created in 1992, the Zhangjiang (ZJ) Hi-Tech Park is much less spread out compared to the Zhongguancun (ZGC) Science Park, since its total area covers only twenty-five square kilometers. While Beijing's information technology cluster focuses on software, Internet, and information services, the ZJ Hi-Tech Park specializes in research in biotechnology and semiconductors.[12] With a total of then thousand companies that employ almost 200,000 workers (among which 20 percent are holders of master's and doctoral degrees), the combined revenue generated in the park has reached about US$30 billion. Given that the park still features undeveloped land, it is expected to grow further and attract hundreds of thousands of workers in the near future. Convenient transportation within the park is possible thanks to the train network that has been recently built. When this book's coauthor Matthew E. Kahn toured the park in the summer of 2013, he was struck by how it looks like a modern suburb with an employment cluster but there is are residential areas and no opportunities for leisure. Within the park he saw no children, schools, old people, pets, stores, or restaurants.

In a 1998 piece published in the *Harvard Business Review*, Michael Porter highlighted the key role that the geographic clustering of industries has on boosting productivity and the generation of new ideas. Porter writes, "Cluster development is often particularly vibrant at the intersection of clusters, where insights, skills and technologies from various fields merge, sparking innovation and new businesses."[13]

Shenzhen

Shenzhen is the best example of a Chinese city exclusively built for economic purposes. What started out as a small southern fishing village in the 1970s has become a large metropolis in just thirty years, and this spectacular growth has had a huge impact on its residents' quality of life.

Shenzhen, located only thirty-five kilometers from Hong Kong, is very distinctive because it is one of the fastest-growing cities in the world. Since 1979, with its designation as a special economic zone (SEZ), this area has adopted free market–oriented economic policies, and this has caused a huge influx of Chinese and foreign direct investment. There are few native Shenzhen people; most of the city's residents are people who migrated there because of job opportunities.

Shenzhen has already become the Chinese city with the most parks, and the environment there is quite pleasant.[14] The 2011 World University Summer Games have helped to expand the subway network, with five new metro lines opening just before the games began. Several people we interviewed declared that Shenzhen is one of their most favorite cities in China because of its natural advantages—vegetation and rain—and its comfortable climate.

As China's economic growth has accelerated, quality of life has diverged in the superstar cities relative to the rest of the nation. Aware and concerned about this trend, the central government has sought to introduce new regional growth policies to increase opportunities in lagging regions.

Opportunities for the Middle Class in Second-Tier Cities

In the United States today, progressives such as Paul Krugman bemoan the declining standard of living for the middle class. Labor economists document the stagnant income for the median household (the middle of the distribution) while the wealthiest 1 percent earn a greater share of aggregate income. Thomas Piketty's book *Capital* has garnered broad attention because of the long-run trends he documents for nations; one of his consistent findings is that around the world the share of national income earned by the top 1 percent and even by the top .1 percent is at a record high.

An active research agenda studies trends in spatial inequality as well. Enrico Moretti's book *The New Geography of Jobs* highlights the huge differences across cities in the clustering of the highly educated. As David Brooks has noted in the *New York Times*, "The highly educated people cluster in just a handful of major United States cities. Decade after decade, educated people flock away from Merced, Calif., Yuma, Ariz., Flint, Mich., and Vineland, N.J. In those places, less than 15 percent of the residents have college degrees. They move to Washington, Boston, San Jose, Raleigh-Durham and San Francisco. In those places, nearly 50 percent of the residents have college degrees."[15] As we have discussed earlier in this chapter, China is now experiencing the same phenomenon with the rise of its high-tech clusters such as the ZGC and ZJ science parks.

The Chinese central government is committed to investing in regional balance and encouraging urban growth and development in less-developed areas, and there are at least four significant place-based policies that are intended to bolster the growth of such regions. The Western Development Program (WDP), launched in 1999, gives infrastructure aid and support for industrial adjustment to western and inland provinces; the program attempts to help heavy industry and the defense industry convert to consumer goods production.

Recent research investigates the ongoing urban transformation in western China taking place between 1988 and 2009, focusing on Chengdu, Kunming, Urumqi, and Xi'an; using land-change maps and satellite images, the research documents growth in all four cities. Each of these cities grew at annual rates near 2 percent from 1988 to 2000, but this annual growth accelerated, to between 5 and 7 percent, after 2006. Each city has more than doubled in size, and nearly one-third of new urban land is outside the core, in what were once small towns.[16]

Some cities in western China are building development zones in order to attract business. The Xining Economic and Technological Development Zone in Qinghai Province is a good example. By 2012, 983 companies had already settled in this development zone. In order to increase educational quality in western China's universities, in 2001 the central government implemented "counterpart support policies" aimed at reducing inequalities in education levels by promoting the development of such universities. Recently, the Guizhou Institute of Technology was built in southwest China, and its first president is a professor from Tsinghua University—China's top university. Western China's new universities are more likely to be successful if quality of life is high in their host cities; such quality of life will allow them to attract better faculty and students. Improvements in air travel, bullet train travel, and Internet access in China guarantee that residents of western cities will not be cut off from access to the fast-paced eastern cities. Yet whether the WDP has a real long-term effect is still an open question; some argue that it has chiefly brought in massive public investment, which has led to a significant improvement in transportation infrastructure and public facilities, but that there is much less private investment.

The central government is also making major investments to foster the development of China's northeast. In the past, cities in this region such as Heilongjiang, Jilin, and Liaoning (the latter two of which are both very close to North Korea), benefited from the emphasis on heavy industry during the Mao years. Since then, like the US Rust Belt, this region has struggled with high unemployment, aging industry and infrastructure, and social welfare bills. While the WDP targets both urban and rural areas, the Northeast Revitalization Program focuses on reinventing the declining cities by developing modern manufacturing and service industries.

A third program is titled The Rise of Central China, and aims to support the development of the central provinces. Up until now, this investment program has focused on transportation infrastructure connecting central China cities into the greater network; the Wuhan–Guangzhou and Zhengzhou–Xi'an bullet train lines opened in 2009 and 2010 respectively, and 2012 saw the start of the Beijing–Guangzhou bullet train line, which stops at several cities in central China. This will promote interregional market integration, thus playing an important role in the development of these cities.

The fourth regional development program targets the "coordinated development" of the Beijing-Hebei-Tianjin region. The three subregions were not integrated well in the past, and Hebei and Tianjin lagged behind Beijing. The new program intends to expedite the development of this northern megaregion to catch up to the Yangtze and Pearl River delta regions in the south. The Beijing-Hebei-Tianjin region has traditionally been involved in heavy manufacturing; under the strategy of regional integration, the area is now becoming a significant growth cluster for automobile, electronics, and high-tech industries, and it is also attracting foreign investments in light manufacturing and health services.[17]

By investing in bullet trains and new highways, the Chinese government seeks to build a system of cities with Beijing as the major hub that will be home to roughly 130 million people.[18] The *New York Times* reports that the new region will link the research facilities and creative culture of Beijing to the port city of Tianjin and the hinterlands of Hebei Province. To accelerate this regional development process and to relieve some congestion in Beijing, the central government is moving many government ministries out of Beijing and relocating them in Tongzhou;[19] this suburbanization of government activity is likely to lead to a clustering of new economic activity at this perirpheral location.

In the United States, the rise of the Washington, DC, metropolitan area highlights the synergies between public government functions and private economic activity as for-profit firms seek to be "in the loop" and to lobby government officials. Research in urban economics has documented that the widespread access to information technology allows firms to fragment, keeping their deal makers located in the high-rent downtown areas and sending their back-office workers to less-expensive areas. In the past such activity had to take place "under one roof" to allow management to be able to cheaply monitor and communicate with different workers within the firm. This same logic also applies to government activity.[20]

The major investment in developing China's poor regions stands in contrast to the standard policy advice offered by Western economists, which is to invest resources directly in poor people.

Nobel laureate James Heckman has stressed the high rate of return for investing in prekindergarten education for all children because such investments can help shape their character and enable lifetime learning. In the case of China, the central government's investment in lower-income regions could have the perverse effect of encouraging people to continue to live in less productive regions rather than moving to other, more productive, areas. If poor cities are inherently less productive (perhaps due to geographic factors, such as being landlocked, or to weak local institutions), investing in such areas may yield a relatively low rate of return relative to investing in areas with better fundamentals or investing directly in initiatives to help poor people. The Beijing example presented above suggests an alternative view: that the growth of cities such as Tongzhou close to Beijing acts as a "safety valve," reducing some of the pollution, congestion, and overall quality-of-life costs that Beijing is suffering from as it continues to grow. If China's urbanites have a menu of viable cities to choose from when deciding where to live, this reduces the likelihood that any one city's quality of life could sharply decline because mobile people would move away from it, and this would relieve some of stress regarding quality of life.

Quality-of-Life Opportunities and Challenges
in Four Second–Tier Cities

Xiamen

Xiamen is located on the southeast coast of China and on the west bank of the Taiwan Straits, with a land area of 1,573 square kilometers; it is a medium-size city with a population of about 3.7 million. It enjoys mild winters and cool summers with an annual mean temperature of twenty-one degrees Centigrade. In 2012 Xiamen's average concentration of particulate matter up to ten micrometers in size (PM_{10}) was fifty-six milligrams per cubic meter—much lower than that in Beijing (at 109 milligrams per cubic meter). Chinese people always think of Xiamen as a beautiful tourist city, and call it the Garden on the Sea, and it has won the "UN-Habitat Scroll of Honour Award. Xiamen boasts abundant tourism resources. When people travel to the city, most of them will visit Gulangyu Island, which is the crown jewel of Xiamen and is known for its variety of international architectural styles and is often called Music Island.[21]

Besides rich tourism resources, Xiamen also has a booming economy. One hundred years ago it was a small fishing village. It was designated as an SEZ in 1980, one of the first in China, and today it has grown into a vibrant urban economic center with per capita gross domestic product (GDP) reaching US$13,200 and an average annual GDP growth rate of 17 percent. It is a major trading port as well as a major destination for foreign direct investment.

The leaders of Xiamen's city government seek to pursue an urban growth strategy based on the city's amenities. The beach access, clean air, and temperate climate offer it the opportunity to compete for talent and new firms. The city is also known for its high-quality urban services; it offers a health care system covering all residents and free, compulsory basic education. For the past three years Xiamen has been listed among the top three out of China's top ten best service-oriented local governments, an honor that recognizes its transparent and efficient government service.

The high quality of life and the rising urban economy in Xiamen are not a "free lunch," however. The average price for newly built homes is 14,500 yuan per square meter in 2013. This means that if someone wants to buy a two-bedroom condominium of roughly one hundred square meters in Xiamen, he or she needs 1.45 million yuan (US$240,000). Xiamen is among the top five "most expensive" cities in terms of housing price in China, just after Shenzhen, Beijing, Shanghai, and Hangzhou.

Chengdu

Chengdu is an ancient cultural city, and as such it can be seen as one of the most influential cities in southwest China. Chinese people always describe Chengdu as a "leisure" city with less pressure and a high quality of life. When you go to Chengdu, you experience a very warm atmosphere that is typical of China's southern cities. Since the weather is pleasant, many elderly people congregate along the street, chatting, watching passersby, or playing games in one of the numerous parks. People in Chengdu seem relaxed, and they enjoy their lives.

The pace of the city is much slower than that of Beijing or Shanghai, and its popularity demonstrates that higher income does not necessarily promise a higher quality of life.[22] For over two thousand years Chengdu's fertile land, abundant rain, and other resources have helped to shape its unique leisure culture.

One can easily observe the leisure culture in Chengdu's teahouses, which are scattered throughout the city. Much more than places to drink tea, they are sites for social networking and places to conduct business, meet with old friends, or make new ones. Chengdu residents love playing mah-jongg, and after work the sounds of mah-jongg tiles can be heard clicking all around the city. Locals will play the game all night, and sometimes even into the next day.[23] Chengdu is also famous for its local cuisine, one of China's four most famous food styles. Known as Chuan-style food, it is noteworthy for its spice and distinctiveness; each dish is unique and flavorful.

Lanzhou

Lanzhou is an industrial city situated in western China, on the shores of the Yellow River and surrounded by mountains. It is one of the most polluted cities in China because the central government made Lanzhou the center for the petrochemical and nuclear power industries and because it is located in a mountain basin.[24] Air pollution is so serious that it is sometimes impossible to see Lanshan, the mountain rising along the south of the city. Ms. Wang holds a doctoral degree from Tsinghua University, and is working for the National Development and Reform Commission in Beijing. Her in-laws are both retired and they settled in Lanzhou many years ago. "We have lived such a long time in Lanzhou that we have already got used to the air quality here," they say. Lanzhou is home to 2.2 million inhabitants, but the city's low quality of life has hurt its attempts to attract high-skilled workers. Some residents leave this city to move to bigger cities. Over the past decade, the population has increased by merely 10 percent, while Beijing, Chengdu, Shanghai, and Shenzhen have seen their populations rise by 35–45 percent.

Urban economists have noted a similar pattern in other nations. Walker Hanlon at the University of California–Los Angeles has explored British city population growth over the years 1850–1910. In those British cities whose industries extensively used coal, particulate levels increased. Such a rise in local disamenities meant that other industries sought to avoid such areas because they could only produce there if they paid high wages as a form of "combat pay" to compensate workers for the low quality of life. In this sense, the growth of one dirty industry can inhibit the growth of cleaner industries because such firms seek out lower cost places to produce.[25]

Ms. Wang's husband lives with her in Beijing because of job opportunities there. But his parents deem that, compared to Lanzhou, Beijing is too big, with too much traffic and high real estate prices. "We prefer to live in Lanzhou because living costs are low, and the city is small, so traffic is convenient," they say. This example highlights a type of comparative advantage such that different types of people will have different rankings of city quality of life. More educated and wealthier people will still be more likely to avoid living and working in cities such as Lanzhou.

Generally speaking, Ms. Wang and her husband believe that Lanzhou is moving in the right direction, and that living standards are constantly rising. More recently, a huge "mountain-moving project" has been started by one of China's biggest construction firms, which plans to flatten seven hundred mountains in order to build a "new metropolis" in the middle of the desert.[26] This new area will provide residents with a whole set of public service facilities, such as schools and hospitals, as well as many job opportunities. Some optimists expect that it will also let fresh air enter into the basin where Lanzhou is located. This example highlights the Chinese central government's willingness to invest in major engineering projects as an means of improving quality of life.

Shenyang

Shenyang, with 5.7 million inhabitants, is a leading example of a major "brown" city. It is the capital of Liaoning Province and the cultural, transportation, and industrial center of northeast China. Heavy industry is the pillar of Shenyang's economy, with more than 50 percent of the city's total GDP coming from manufacturing industries. Ms. Liu and her family left their small rural town to settle in Shenyang about seven years ago because she wanted her daughter to grow up in the city and receive a better education. She was not used to living in a big city, and when she first arrived urban construction had just started. She needed to walk far to reach a bus station, and most of her time was spent commuting.

Rapid growth has created serious pollution problems in Shenyang due to pollution from transportation and industry. "But at least I was satisfied because they had constructed a bus station close to my house," Ms. Liu says. The local government started to focus on recruiting new industries that cause less environmental damage, and it succeeded in improving air and water quality, even if there are still some occasional problems. It is said that the water in Ms. Liu's district contains excessive amount of heavy metals. Most people have already installed water purifiers in their homes, but Ms. Liu, a chemistry teacher, thinks that these purifiers are not able to protect against heavy metals.

This story reveals that Shenyang residents do care about their quality of life, and they also demand higher environmental regulation. But when we ask Ms. Liu about Shenyang's heavy industries, she answers that Shenyang needs those factories: "Heavy industries will bring more job opportunities for my students." Most graduates working in Shenyang say that they stay there because of the low housing price (the average monthly rent for a thirty-square-meter apartment is five hundred yuan).

China's Emerging System of Cities

As China relaxes its regulations on internal passports (the *hukou* system), its people will be able to choose among a wide list of cities as potential places to live. While such households do not directly vote for local leaders, they do have the freedom to "vote with their feet" and migrate to a city that offers the economic opportunity and the quality of life they desire and can afford.

The ability to move across communities and cities provides a type of insurance policy for urban workers. If a specific city's quality of life declines, a household can move elsewhere. Such a move would cause the sacrifice of an existing social network, and would bring a moving cost, but those relocating could re-create their lives in a new city. Improvements in information and access to microblogs provide households with real-time information about opportunities and challenges in different cities.

In late July 2014 China's government announced that it would reform its *hukou* system that has traditionally inhibited large-scale migration from the country's rural areas to its urban centers. The announcement notes that "the government will remove the limits on *hukou* registration in townships and small cities, relax restrictions in medium-sized cities, and set qualifications for registration in big cities;" this means "there will be no limits for settlers moving to small cities." Additionally, "Medium cities, defined as cities with populations between one and three million, will have a low barrier to entry. Large megacities, defined as cities with over 5 million residents, will still have a fair amount of restrictions and potential residents will have to qualify under a 'points system' that takes into consideration a variety of individual factors including 'seniority in employment, their accommodation and social security.'"[27] Relaxing the *hukou* migration restrictions in smaller cities first means that the state prefers a more "balanced" population distribution across small-, medium-, and large-size cities.[28] If the reform had relaxed the *hukou* constraints in all of China's cities, it is likelier that most rural migrants would have preferred larger, more established metropolises, and Beijing and Shanghai would have faced more congestion and pollution.

In a June 2014 speech given on behalf of the International Institute for Strategic Studies in London, China's premier Li Keqiang noted,

Due to inadequate financial resources and limited public goods, the 200 million rural migrants cannot enjoy the same public services as those who have been living in cities with urban residency. Between now and 2020, we need to grant urban residency to 100 million of such rural migrants, mainly in eastern China. We also need to encourage another 100 million rural residents in central and western regions to settle down in cities and towns nearby. In addition, there are 100 million Chinese living in run-down city areas, which need to be redeveloped at a faster pace. To address these three problems, each involving 100 million people, is a pressing task for China as it pushes forward the new type of urbanization.[29]

By investing in intercity transportation infrastructure, China's central government has increased the menu of possible urban destinations for workers and firms. Over the last twenty years the nation has made enormous investments in highways and airports. These investments help to integrate the nation's cities, facilitating trade, travel, and learning. In 1990 there were only seven airports that could accommodate Boeing 747s, and the nation's entire highway system spanned less than three hundred kilometers. By 2006, however, roughly 70 percent of the cities had been connected by highways and the airline network; the average annual growth rate of the highway system was 52.6 percent in 1995–99 and 20.4 percent in 2000–2004.

Bullet Trains Connect China's Cities

In recent years the cities of Beijing, Guangzhou, and Shanghai have experienced enormous real estate price growth. High real estate prices reduce middle-class quality of life and inhibit young people from starting their own families. China's new bullet trains will offer young people the option of living in less-expensive cities while still having increased access to nearby superstar cities.

The introduction of the bullet trains gives firms the option of locating their headquarters in the major cities and send other activity to less-expensive cities nearby.[30] Firms that need infrequent access to the major city's deal makers and government officials can decentralize and locate in the bullet-train-accessible second- and third-tier cities.

The introduction of bullet trains creates the possibility that middle-class Chinese can have "the best of both worlds": access to less-expensive housing while still being able to access the productive and politically powerful superstar cities. In 2006 there was no bullet train in China, but by the end of 2011 there were fifteen major bullet train lines and about three thousand kilometers of upgraded railway lines in operation with speeds of 200 kilometers per hour or more; at the end of 2012 the nation had the world's longest bullet train network, with more than 9,300 kilometers of track in service. About a quarter of Chinese cities at the prefecture level or higher been connected by bullet train lines.[31]

In the early 1990s most Chinese conventional trains moved at speeds below sixty kilometers per hour. This speed was raised several times in the late 1990s and early years of the twenty-first century, but even then it did not exceed 150 kilometers per hour. The Ministry of Railway (MOR) announced its ambitious bullet train plan in 2006. The typical financing arrangement for constructing bullet train lines is that the MOR pays 50–60 percent of the total cost and the destination cities pay the other 40–50 percent. The first group of bullet train lines was opened in April 2007, boosting the speed of some major trains as high as 200–250 kilometers per hour. In August 2008, new bullet trains between Beijing and Tianjin reached a higher top speed of about 350 kilometers per hour. By the end of 2010, China's bullet train service length reached 8,358 kilometers, which was a distance record for the world. The MOR plans to build a "four plus four" national passenger designated line grid (four north–south and four east–west high-speed rail corridors) by 2020. At that time the total service high-speed rail railway system will reach an overall distance of twelve thousand kilometers.

China's enormous investments in bullet trains contributes to creating an integrated system of cities that is based on a low carbon technology (bullet trains connected to urban subways) rather than on intercity highways used by private cars. This investment is likely to have significant benefits for second- and third-tier cities connected to the first-tier cities that are too far to reach by car but too close to travel to by airplane.

In figure 3.1 we provide a geographical overview focused on which

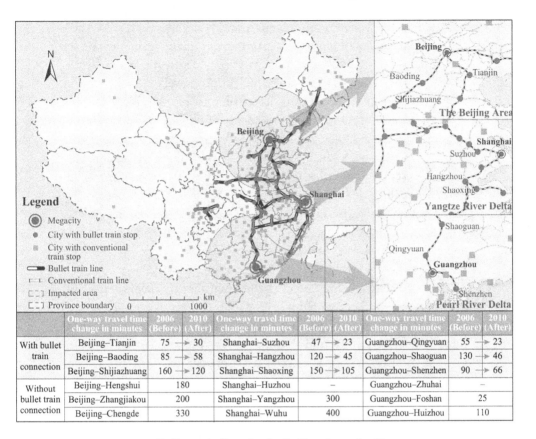

	One-way travel time change in minutes	2006 (Before)	2010 (After)	One-way travel time change in minutes	2006 (Before)	2010 (After)	One-way travel time change in minutes	2006 (Before)	2010 (After)
With bullet train connection	Beijing–Tianjin	75 →	30	Shanghai–Suzhou	47 →	23	Guangzhou–Qingyuan	55 →	23
	Beijing–Baoding	85 →	58	Shanghai–Hangzhou	120 →	45	Guangzhou–Shaoguan	130 →	46
	Beijing–Shijiazhuang	160 →	120	Shanghai–Shaoxing	150 →	105	Guangzhou–Shenzhen	90 →	66
Without bullet train connection	Beijing–Hengshui	180		Shanghai–Huzhou	–		Guangzhou–Zhuhai	–	
	Beijing–Zhangjiakou	200		Shanghai–Yangzhou	300		Guangzhou–Foshan	25	
	Beijing–Chengde	330		Shanghai–Wuhu	400		Guangzhou–Huizhou	110	

Figure 3.1 Bullet train lines in the Beijing Area, the Yangtze
River delta, and the Pearl River delta in 2010

Source: Siqi Zheng and Matthew E. Kahn, "China's Bullet Trains Facilitate Market Integration and
Mitigate the Cost of Mega City Growth," *Proceedings of the National Academy of Sciences
of the United States of America* 110, no. 14 (2013): E1248–53.

cities are located close, but not very close, to the three megacities of Beijing, Guangzhou, and Shanghai. Tianjin is 120 kilometers from Beijing,
and there is no flight between them. The Beijing–Tianjin bullet train ships
420,000 passengers per week, and substitutes for conventional train, car,
or plane travel depending on the distance. Cities located 80–250 kilometers from the megacities are far enough from them to not be easily ac-
cessed by car yet close enough to make flying impractical. The one-way price

of a ride on the Beijing-Tianjin bullet train is fifty-five yuan (US$8.90), roughly two and a half times the conventional train fare, and the fare for the Wuhan–Guangzhou bullet train is about 60 percent of the airfare. China's middle class can easily afford bullet train travel.

Improvements in transportation infrastructure between cities offers one strategy for mitigating the megacity quality-of-life challenge. China's recent investment in bullet trains allows individuals to move at speeds of roughly 175 miles per hour, and this increases the number of locations that have access to the megacities. If individuals can swiftly move from smaller cities where they live to megacities, they can enjoy the benefits of megacity access without suffering the social costs associated with megacity growth. Workers and firms who need to attend face-to-face meetings with firms and government officials in the megacity once a week or once a month can have a lower cost of living by relocating to a "satellite city."

The introduction of the Chinese bullet train alleviates some of the congestion costs associated with urban growth in the megacities while simultaneously increasing the labor demand and human capital levels in the now more accessible second- and third-tier cities. The smaller nearby cities enjoy an increase in their access to a larger market for goods, and a larger number of potential employers are accessible. The cultural and consumer attributes of the big city are now in reach for these satellite cities. Consider the travel time from Tianjin to Beijing. This journey would take one and a half hours by private car (and vehicle ownership rates remain low in China) but only a half hour by bullet train. In this sense the bullet train creates the possibility that the second- and third-tier cities nearby become a safety valve for the megacity, and this alleviates concern about such a city growing "too big." High home prices and relatively low quality of life in the megacity will encourage some households and firms to consider relocating to the second-tier city.

Great Expectations?

The global interest in Tomas Piketty's work on household income inequality highlights concerns that the quality of life of the urban middle class and poor has stagnated. The people of China are well aware of the great strides the country has made in recent decades, but an ongoing debate focuses on the quality of life of the poor as overall income inequality in China rises.

When the poor move to China's cities, what do they expect to gain from this migration? Political stability hinges on whether their expectations are met. The political challenge in urbanizing China is whether the new migrants to cities expected even greater progress in their quality of life than they have actually enjoyed. Has Ms. Fu achieved her goals? Is she optimistic about her children's future? Migrant workers like Ms. Fu do have much higher earnings compared to their counterparts who stay in rural villages, but they do not enjoy many of the benefits of urban life. They are sometimes subjected to labor market discrimination, social segregation, and status inferiority. Constrained by the *hukou* system, Ms. Fu is not eligible for low-cost public housing units, health insurance, unemployment insurance, or a pension fund in Beijing, and she cannot send her two kids to any public schools there. She does not have many local friends except the family of this book's coauthor, Siqi Zheng. On weekends she stays in her small room, watching TV and chatting with her two sons and her aging parents on the phone.

The good news is that the central government has decided to remove the *hukou* limits in small- and medium-size cities; migrant workers in those cities will soon enjoy the same benefits that local people do. But it is still very difficult for those low-skilled migrants to obtain *hukou* in large cities such as Beijing and Shanghai. Therefore, they have a strong incentive to move from the megacities to smaller cities, perhaps those closer to their home villages. This is encouraged in Premier Li's three-hundred-million-person migration policy package. Members of the middle class will have more choices, because it will be easy for them to qualify under the "points system" to receive a big city's *hukou*. But they will also face the challenge of being squeezed out by those in the top 1 percent of the income distribution who bid up housing prices and living costs in the superstar cities.

Every would-be urban dweller must choose what city to live in and where within that city to live. In chapter 4 we turn to exploring the trade-offs embedded in intercity locational choice.

CHAPTER FOUR

The Causes and Consequences
of Chinese Suburbanization

For years, US basketball player Michael Jordan has owned an enormous home in suburban Highland Park, Illinois, twenty-eight miles north of downtown Chicago. The home, built in 1995, features nine bedrooms, fifteen bathrooms, its own indoor basketball court, and five fireplaces, among many other amenities; it has a garage that can hold fifteen cars and has 32,683 square feet of interior space. While Jordan shares an American Dream similar to that of millions of other Americans, he lives on a much grander scale than most.

In contrast, China's cities are currently much more compact, and few people live in single-family homes. In Chinese cities in the year 2007, the average size of a dwelling unit was eighty-five square meters, and only 2.8 percent of urban households live in detached single-family homes. Over 80 percent of urban dwellers live in middle-rise and high-rise residential buildings.

As Chinese urbanites grow wealthier, will they "be like Mike"? Will China's cities sprawl? This chapter examines the private benefits for such households and the social cost from this lifestyle choice. We first discuss the current situation in urban China with respect to where households and firms locate. We then discuss some emerging trends and conclude by discussing the challenges that the state and central government faces due to these trends. The economic geography of economic activity within China's cities matters because China's urbanization is playing out right now. Given that compact, high-density cities tend to have a smaller carbon footprint than more dispersed cities with a larger land area, it is important to understand the private demand on the part of households and firms concerning where they want to locate within Chinese cities.[1]

A Case Study of Beijing's Growth

Beijing's urbanized area grew from 346 square kilometers in 1980 to 1289.3 square kilometers in 2010. When coauthor Siqi Zheng was an undergraduate student at Tsinghua University in the late 1990s, the fourth ring road (with a radius of ten kilometers) was under construction, and Beijing only had two subway lines. Today the sixth ring road (with a radius of over twenty-five kilometers) is now open and there are sixteen subway lines in operation. Figure 4.1 shows how Beijing has expanded over time.

We seek to explain the geographic distribution of jobs and people within China's major cities and how these patterns are likely to change over time. With an understanding of these trends, we can make informed predictions about how different indicators of pollution and "greenness" are likely to evolve.

Beijing's central business district (CBD) has greatly expanded, and residential land use has extended into the suburbs. Industrial land use has been pushed out from the city center. Built-up urban areas have quickly expanded, and new mass housing complexes have been built around the rapidly expanding urban fringe. When Siqi was an undergraduate student at Tsinghua University fifteen years ago, there was plenty of farm land close to the university's east entrance. Today this area has been converted into a very busy science park and a vibrant urban neighborhood with dozens of high-rise office buildings, fancy restaurants and coffee shops, and high-end condominium complexes.

One major trend taking place in Beijing is the relocation of the large, manufacturing-based state-owned enterprises (SOEs) from the central city to remote suburban areas and nearby Hebei Province. During the prereform socialist era, these SOEs often occupied good locations in the cities. At the end of 1999, the Beijing municipal government announced its plan for SOE relocation: 738 manufacturing firms within the fourth ring road would be relocated, either to the sixth ring road or to second-tier cities

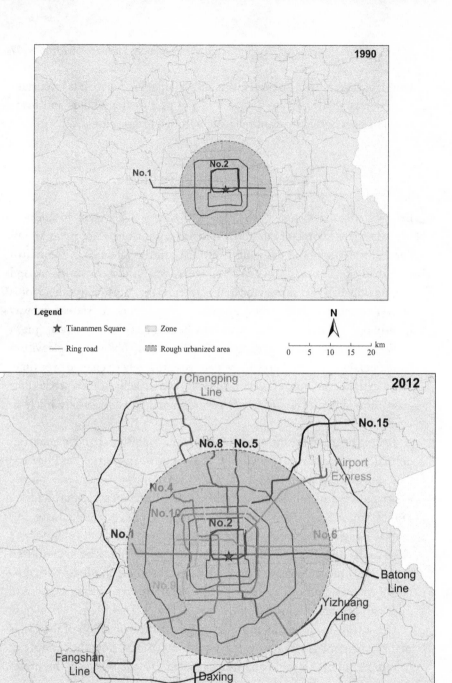

Legend

★ Tiananmen Square   ▨ Zone

—— Ring road   ▨ Rough urbanized area

N

km
0   5   10   15   20

Legend

★ Tiananmen Square   ▨ Zone

—— Ring road   ▨ Rough urbanized area

N

km
0   5   10   15   20

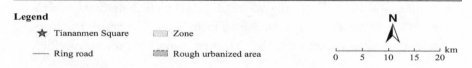

Figure 4.1 Expansion in Beijing

near Beijing, depending on their type of industry. The Beijing No.1 Machine Tool Plant's original site was in Beijing's CBD, in Chaoyang District. On May 18, 2000, the real estate developer Hongshi signed a contract with the plant and paid it a considerable amount as relocation compensation. The plant moved to the Shunyi District outside the sixth ring road. Hongshi also paid the Beijing local land authority the land transfer royalty to obtain the right to develop; the developer then built a luxury residential complex on that site.

Since the reforms introduced in the early 1980s, China's cities have experienced a divergence in economic outcomes as Beijing, Guangzhou, Shanghai, and Shenzhen have boomed. Within those cities innovative clusters have emerged. One example is the information technology (IT) cluster of the Zhongguancun (ZGC) Science Park in Beijing, located in the core area very close to Peking University, Tsinghua University, and the Chinese Academy of Science.

Twenty-five years ago this area was full of old one- and two-story residential buildings. After he visited Silicon Valley, Chunxian Chen founded the Advanced Technology Service Division in ZGC in 1978. Throughout the 1980s, this area was known as Electronics Avenue because of the large number of stores that sold electronic components and devices. Later, with the booming IT industry, this area took advantage of its strategic location near top universities and research centers to attract highly educated people and to transform the ZGC area into a center of technological research and development.

In 1988, the central government officially approved the ZGC Science Park as China's first high-tech innovation center. Thanks to tax exemptions and research subsidies, the ZGC Science Park has attracted thousands of Chinese and international companies such as Baidu, Google, Hewlett-Packard, IBM, Intel, Lenovo, Microsoft, Nokia, Oracle, Samsung, and Siemens. Today the park covers a total area of almost 500 square kilometers (equivalent in area to the city of Prague) in the northwest of Beijing. The ZGC Science Park is home to twenty thousand high-tech companies and has generated combined revenue totaling US$386 billion in 2012, with an average 20 percent increase for five consecutive years. More than one million people work in this area, which attracts highly educated workers from China and abroad.

Figure 4.2 The ZGC area in Beijing

In addition to a steady increase in the number of top global technology-oriented enterprises, the ZGC area features diverse shopping and retail stores that boosts its quality of life. In recent years, shopping malls and restaurants have been flourishing, and a wide variety of entertainment services are now provided (see fig. 4.2). A snowball effect has taken place as the clustering of a large number of sophisticated wealthy consumers in one geographic area creates new business opportunities for restaurants and retailers who cater to them. As such upscale stores cluster in these locations, areas such as ZGC become even more desirable as they offer economic opportunities and offer the benefits of a major "consumer city." In such a setting, environmental quality will be highly valued because outdoor leisure time will be more enjoyable due to lower air pollution levels.

Some weekends Siqi sits in a Starbucks shop in ZGC to have a coffee. She always finds that in the Starbucks young people are talking about how to establish their own IT company and attract venture capital investment. They are full of energy, hope, and innovative ideas. For decades, Western companies have accused Chinese firms of pirating software and ideas, but these new clusters of innovation raise the likelihood that China will soon be a major producer of its own intellectual property.

The long-run success of clusters such as the ZGC Science Park hinge on these areas offering a high quality of life because high-skilled workers have many choices and are footloose. The Western media continues to publish stories about highly educated Chinese adults seeking out opportunities to live and work in Australia, Canada, and the United States. Such a brain drain will be less likely to take place if quality of life improves in China's major cities.

Chinese Urban Housing Demand Today

China's National Bureau of Statistics regularly surveys the population to learn about its living standards. In 2007 a large-sample urban household survey revealed that the average area of a residential unit is 84.5 square meters, and that 82.3 percent of Chinese urban households owned their homes. This share is much higher than that in US cities (at about 67 percent). The median age of a condominium buyer in Beijing is thirty-two.

Wealthier people are more likely to own a condominium, and 87 percent of those in the top 10 percent of the income distribution own their homes, but this share drops to 73 percent for the poorest 10 percent (which, it should be noted, is still much higher than the same statistic for poor people in the United States). In Chinese cities, wealthier people live in larger homes. In 2007, the poorest 10 percent occupies 67.8 square meter of floor area per household, on average, but the wealthiest 10 percent consume about 107.3 square meters per household. Roughly 15 percent of Chinese urban households own a second home, and this share is much higher in the first-tier cities.

In 2007 about one-third of all existing housing units were erected by real estate developers in the form of large residential complexes. Another one-third are privatized public housing units as a legacy from the old housing regime prior to the 1990s. During the earlier communist era, urban housing in China was allocated to urban residents by their employer (the work unit) through the central planning system. Workers enjoyed different levels of housing types depending on their office rank, occupational status, and level of work experience. Governments and work units were responsible for housing construction, and residential land was allocated through the central planning system. Workers would also obtain services, such as child care, education, and family health care, from their work units. Larger and more prominent work units would provide better housing. Residents had little opportunity to choose their residential location in the absence of a housing market. Since the 1980s, most of the work-based housing units have been privatized and sold at a small percentage of their true market value (less than 10 percent). Such subsidies implicitly transferred income to existing urban households. By the end of the 1990s, housing procurement by work units for their employees had officially ended, and new homes were built and sold by real estate developers in the market.

The Demand for Living in the City Center

Every urban family must choose where it will live and work within a specific city. Such households face a budget constraint, and housing is a major budget item. Housing prices vary within a city such that less desirable properties will be less expensive to purchase. The market price for housing confronts households with making a series of trade-offs. Only the very wealthy can live in a large housing unit close to key amenities.

Nobody likes to waste time commuting. If jobs and the city's main cultural attractions are located downtown, people will want to live closer to downtown. Such demand for limited downtown space bids up the price of land, and this encourages developers to build skyscrapers that economize on that land. Those plots of land that are farther from the city center will sell for a price discount to compensate workers for the lost time in commuting. The benefit of living farther from the center is that land is less expensive and there are fewer people.

In the case of China, jobs are centralized downtown. In the Beijing metropolitan area's "core eight" districts (Chaoyang, Chongwen, Dongcheng, Fengtai, Haidian, Shijingshan, Xicheng, and Xuanwu) in 2004, 23.7 percent of all jobs were located within three miles of the city center (calculated by the authors using 2004 economic census data). High-quality public services are also centralized; more than 70 percent of key primary schools and grade-A hospitals are located within the third ring road in Beijing, while only 30 percent of Beijing's population live in this central area. The fact that the best schools are located close to the city center causes traffic congestion. Roughly half of the students in Beijing are picked up from and taken to school by their parents, and roughly 40 percent of those parents drive. A recent study by Ming Lu, Cong Sun, and Siqi Zheng documents that, all else being equal, the traffic congestion during the morning peak on school days is 20–30 percent higher than on school holidays.[2] This vehicle idling contributes to the city's air pollution problem, significantly raising pollution concentrations on a typical school day. This problem could be mitigated if real estate developers built high quality schools close to their new apartment towers.

The wealthy in Beijing live downtown. Their willingness to pay for downtown locations is significantly higher than it is for the poor because most job opportunities and good public services are concentrated in the central city. Given China's one-child policy and the fact that educated people generally marry other educated people, China's major cities are now home to an increasing number of "power couples" in which both spouses work at high-paying jobs. If these spouses work in different places, the couple must decide who will bear the brunt of the commuting. Siqi works at Tsinghua University in northwest Beijing, fifteen kilometers from the CBD, and her husband works in a real estate investment company in the CBD. Since Siqi's job is more stable and Siqi's son will attend primary school and high school on the Tsinghua campus, they chose to buy their house near the Tsinghua campus.

For "power couples" time is valued very highly, and they are thus concerned about commute times. If both husband and wife work in the center city, residential neighborhoods near there are especially desirable even if they are expensive. Similarly, married couples without children, who are often more career oriented, have a preference for living close to the center city. In contrast, senior workers over age fifty prefer less central locations, as they are going to retire and want to enjoy a more relaxed life with more open space and less noise.

Where Are the Consumer Centers in the Metropolitan Area?

US urban economists have argued that cities that are fun to live in will be more likely to grow in the future.[3] The logic underlying the "consumer city" theory is that this is a footloose age. Companies such as Facebook can locate anywhere, but the scarce resource for such firms are the people

who might work for them, and attracting and retaining talent requires locating in a desirable area. In this sense, places with a high quality of life
become productive places because they are desirable, able to attract and
retain high-skilled individuals. An example of this phenomena in the
United States is the rise of Silicon Beach in the Los Angeles coastal areas
of Santa Monica and Playa Vista. Leading tech companies such as Google
and Yahoo are building smaller corporate campuses in such high-amenity
areas.

Once the high-skilled cluster is in a specific geographic area, the area's
quality of life is likely to improve. Spatial clusters of educated, high-income
individuals attract better restaurants, shops, and cultural institutions.[4]
After all, Starbucks does not throw darts at maps when it decides where to
open a store; it performs a market demographic analysis to identify areas
where there is sufficient demand such that a new store could turn a profit.

Our own research shows that as Beijing municipal government invested in the 2008 Olympic Village and in new subways, local home prices
increased, developers constructed more new housing nearby, and more
restaurants of higher quality opened. Households with a higher income
and better education were attracted to living in those places.[5]

The restaurant industry provides a good indicator of gentrification and
the rise of the "consumer city."[6] If more people are moving into an area, and
if they are wealthier than the residents already living there, we can expect
to see more fancy restaurants to come to those areas. In our research project
we have identified the restaurant chains that fit the preferences (taste, service quality, and price) of the upper middle class and the wealthy who can
afford new commodity housing. In China, Starbucks seeks to locate where
wealthy urbanites live and work. When coauthor Matthew E. Kahn first
visited Shanghai in the summer of 2013, he went to a Starbucks shop and
asked a worker how many shops Starbucks had in Shanghai. He guessed
that there were fifteen; the worker corrected him, telling him there were
250!

Figure 4.3a shows restaurant density in four-square-kilometer cells in
Beijing. We can see that this restaurant density is correlated with popula-
tion density, but restaurants are more concentrated in the central area.
Figure 4.3b shows where Starbucks coffee shops are located in Beijing.
There are more than two hundred Starbucks shops in Beijing, and they

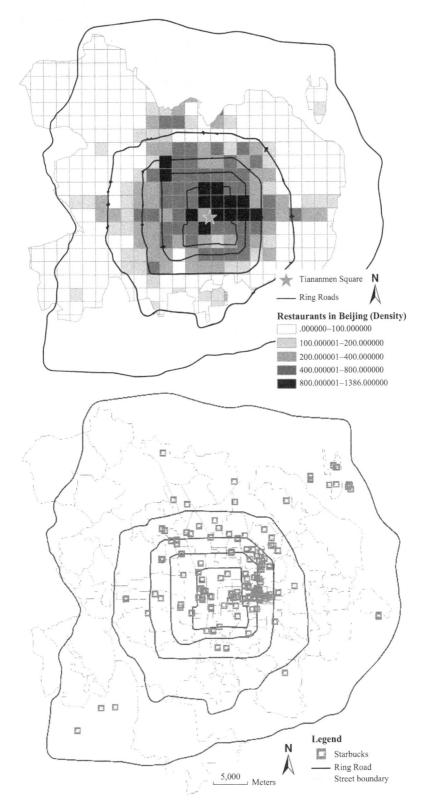

Figure 4.3 Restaurants and Starbucks coffee shops in Beijing

have clear clusters where talented people work and live: the CBD (Jian-guomenwai), the financial district (Jinrongjie), the IT subcenter (Zhong-guancun), the large residential area north of the CBD (Wangjing), the Beijing Capital Airport, and the area close to Tsinghua University and Pe-king University.

The Urban Poor's Housing Demand

There are two groups of the urban poor in Chinese cities. The first group is that of the local poor people with *hukou* internal passports. Many of them own old houses in good locations as a legacy from the former cen-tral planning era. These old houses are small and of low quality relative to the modern building stock. Developers seek to buy this land and demolish the existing homes in order to build new towers. The local poor who own those houses receive a windfall if developers purchase their old houses at a market price, but in some cases the government seizes their homes with-out compensation. In recent years, premier Li Keqiang has tried to trans-form the slums the local poor people live in (known as *penghuqu*, "shanty areas"). New affordable residential complexes with good neighborhoods and public facilities have been built to accommodate those urban poor, and millions of them have benefited or will benefit from this program. The program has three main components: (1) the urban poor can move to a better living environment; (2) by moving the urban poor into new com-munities, the local governments can obtain more land in the central city where the old slums were located; and (3) the huge construction invest-ment will boost economic growth. Premier Li repeatedly emphasized that the local governments must understand that the main goal of this program is to improve the local poor's quality of life, not securing land or boosting investment. The Chinese central government plans to move ten million poor families out of their old, shabby houses by 2017, especially in the in-dustrial cities.[7]

The second group of the urban poor are the "floating" migrants who spend 90 percent of a year working in cities and then bring their earn-ings home to their rural hometowns during the Chinese New Year. *Cheng-zhongcun* (urban villages) are a typical type of informal housing that rural

Figure 4.4 *Chengzhongcun* in Beijing

migrants live in. Residents there must cope with high crime rates, inadequate infrastructure and services, and poor living conditions (see fig. 4.4). Hundreds of urban villages exist in large cities such as Beiijng, Guangzhou, and Shenzhen. There were 139 urban villages in Guangzhou in 2006, and urban villages in Guangzhou and Shenzhen make up more than 20 and 60 percent of their planned areas, respectively, providing homes to 80 percent of migrants in these cities. Siqi Zheng conducted a survey of those urban villages in Beijing in 2008. Most of the housing units in the surveyed urban villages are one-room units, and the mean living space per dwelling room is 13.2 square meters. The mean per capita living space is 8.2 square meters, only one-third of that found in Beijing's formal housing sector (27 square meters). Facilities in urban village units are inadequate and poorly maintained. More than 90 percent of the surveyed units do not have bathrooms or kitchens; dwellers in these units use public bathrooms and cook in public spaces. Despite the cold winters and hot summers in Beijing (daily minimums in January average 15.1°F, and daily maximums in July 87.4°F), 86 percent of the units Siqi surveyed have no heating and 93.3 percent lack air-conditioning. Eighty-one percent of migrants stated that they had personally observed crimes in their neighborhoods in the six months prior to the survey.

City governments seek to demolish such informal housing, and rural migrants are pushed farther out to the remote suburban areas. But the displaced do not leave the city, because they can find jobs here. In recent years some cities have started to provide public rental housing units to those rural migrants without *hukou*, which is good news for displaced migrant workers. But Beijing still excludes those rural migrants from its public housing program, so the end result is that the displaced resettle in the urban villages in fringe metropolitan locations and thus have a longer commute to their place of employment.

The Housing Supply in Cities

New housing is supplied by real estate developers in the form of large residential complexes. A typical residential complex developed by a single developer usually consists of several high-rise condominium buildings that share nearly the same location attributes, a common architectural design and structure type, and community/property services and contain hundreds or even thousands of housing units. A large complex may be divided into several phases, and those phases are developed and sold sequentially. Each phase contains a couple of high-rise buildings. The housing units within a single building are very similar, but differ with respect to attributes such as number of floors, unit size, number of bedrooms, and the direction the main bedroom faces.

In China's cities, developers can only build new housing towers when there is a government land auction. A necessary first step for building a new housing project is to acquire the use right of a land parcel from the city's land authority. All urban land is owned by the state in China, and starting in 1988, the Chinese government began to offer long-term leases of land parcels.[8] Any party who wants to acquire the land-use right in the pursuit of profit must pay a lease fee as a lump sum at the beginning of the lease period.

Before 2004 the conventional way to lease land was through negotiation between the developer and the local government. After 2004 the Chinese central government required that all land leases be privatized though an open-bid auction process. This switch was motivated by corruption concerns during the negotiations that were leading to less revenue being collected by the state. A closed-bid auction creates the possibility of side payments such that a low bidder wins the auction. An open-bid auction is less likely to suffer from this problem if bidders are unable to collude with each other to shade down their bids. In a competitive bidding process, the land will be allocated to the developer who values the scarce resource the most. Land Reservation Centers (LRCs), departments within cities' land authorities, were established to implement land auctions. After acquiring land parcels from rural villages as well as original urban occupants, and removing the previous occupants, an LRC engages in such urban land improvements as providing basic road access and connecting the parcel to the electricity and water network. The LRC then puts the land parcels on the market for an open and competitive auction. The developer who bids the highest acquires the parcel, and local government receives the revenue.

City governments control the land supply in their metropolitan areas. The development of the housing market in Chinese cities in the past two decades has been fueled by the dramatic increase in the land supply on the part of city governments. Land sales are a major share of a city's revenue each year. Starting in the late 1990s local governments started to collect land transfer fees in order to finance public goods and services such as roads, subways, sewage systems, education, and health care facilities.

How many units a developer can build on a land parcel is determined both by the size of the land parcel and its zoned density. The density is technically measured by the permitted floor-to-area ratio (FAR, the ratio of floor area to land area) on that parcel. The FAR is predetermined by the land authority, which chiefly considers the land's location value and consults the city of Beijing's master plan. Developers who have strong bargaining power with government agencies have been able to build taller buildings than is usually permitted by FAR.

Our own research yields a surprising finding that the density of new residential buildings does not decline with distance from the city center; nor does the dwelling size change significantly from the city center to the suburbs. This finding would surprise people who have lived in many cities in the United States. Consider flying into any major US city; at the city's center, where land is most expensive, there are enormous skyscrapers, but farther from the city center single-family homes can be found where land is less expensive. While a skyscraper is expensive to build, it economizes on land consumption per person living or working in the building. Basic economic theory predicts that in areas featuring higher land prices developers will seek to build taller buildings.

The urban planning principles specified by the Beijing local government help explain the puzzle of why Beijing's population density is not much higher closer to the city center. Tiananmen Square is a political landmark in China, and the Forbidden City behind it is the most important historical heritage site. To keep these landmarks prominent, Beijing's urban planning commission set rigid restrictions on the height of buildings near Tiananmen Square in the city center. The urban planners also follow another planning principle: as distance from Tiananmen Square and the Forbidden City increases, buildings heights should also go up to create a skyline for Beijing.

Industrial Parks and the Birth of "Edge Cities" in China

Several emerging trends contribute to suburban growth. All around the world, improvements in road networks and information technology allow firms to move toward less-expensive suburban locations. One way to see where job suburbanization is taking place is to drive to cities in Yangtze River delta, such as Suzhou and Wuxi. When you are still far away from the downtown areas, you will already see many factories along the highways.

Another push for job suburbanization is the growth of industrial parks in many suburbs. Local governments are keen to build large industrial parks (special economic zones, or SEZs) on the city fringe with inexpensive land and favorable tax deduction policies to attract foreign direct investment (FDI) and firms that can produce high tax revenues. Following the first development zone established in Dalian in 1984, thousands of industrial parks have been opened across Chinese cities, and their importance has grown rapidly. There were 1,568 industrial parks established in China by the end of 2006. Among them, 222 are nation-level industrial parks approved by the State Council of the People's Republic of China. The remaining 1,346 parks are provincial industrial parks. Based on the data from the National Bureau of Statistics, in 2006 FDI received by the fifty-four national-level industrial parks (or economic development zones) amounted to US$130 million and contributed to 21.6 percent of the total in China; their total value of merchandise exports was US$1,138 million, 15 percent of the total.[9] Most of the industrial parks locate in suburban places in southeastern coastal cities. Those industrial parks that feature high-tech job opportunities, a high-skilled labor force, and a strong infrastructure base become vibrant new towns. The Tianjin Economic-Technological Development Area (TEDA), established in 1984, is one such successful industrial zone. The six key industries in TEDA are electronics, biochemicals, light industries, manufacturing, automobiles, and logistics. Several of the world's top five hundred enterprises locate in this thirty-three-square-kilometer area; it was later renamed the Binhai New Area.

In China industrial parks are an important place-based policy tool for facilitating economic development. A mayor faces a key trade-off when deciding whether to set aside scarce land to build an industrial park; when he does so, he bears the early risk because of the up-front costs of land conversion and infrastructure that are necessary for building the park. He also sacrifices land revenue he would have collected from a land auction to developers. Once the park is built and is occupied in the medium term, productivity gains take place, but these are only enjoyed in the medium and the long term. (The term length of a mayor is only four years, and he can hold this position for a maximum of two terms.) But if the land and real estate markets are efficient, future benefits will be capitalized into current land and real estate values. If people believe that a productive industrial park will effectively trigger residential and commercial development nearby, the land value nearby will appreciate ahead of time. In this way the mayor can capitalize the future benefits of a park into current land-sale revenue. In fact, Chinese mayors are engaged in such "land value capture" by selling nearby land parcels for a price premium. But if a mayor locates an industrial park in the wrong place and the market does not respond, this may cause a major cost overrun because future cash flow of tax and land-sale revenue will be small.

Our own research looks into the role that industrial parks play in shaping major Chinese cities' urban form through creating new employment subcenters and new residential neighborhoods. Physical proximity between firms both within and nearby industrial parks who seek to colocate and agglomerate facilitates local productivity growth. The well-paid workers at these new employment centers seek nearby housing and retail opportunities. Real estate developers and retail entrepreneurs respond to this increased local market potential by supplying new housing and consumer amenities. This chain reaction of new jobs causing real estate construction and the opening of new high-end retail opportunities creates a vibrant "edge city." In the United States, urbanists such as Joel Garreau have documented the rise of such edge cities clustering around suburban airports far from the city center. In the case of China, we have documented a similar process.

Mayors acquire farm land and convert it for more productive use. New industrial parks stimulate economic growth because they solve both coordination problems and land assembly problems. Establishments that seek to agglomerate within a small geographic area face transaction costs in simultaneously seeking to colocate. Even if they could agree to do so, there is no reason to believe that a major city would have a large enough vacant plot of land that would allow them to simultaneously execute this plan. In this sense, the industrial park permits a degree of coordination in a timely fashion that is unimaginable in a setting (think of downtown Manhattan today) featuring a land market with preexisting durable structures and long lags between sales dates.

Using firm-level data, we find that firms in industrial parks enjoy 15 percent productivity premium on average, and the manufacturing job density in those parks is twice that in other places. By geocoding the exact location of already existing firms within the city and by combining this with information on the exact location of new industrial parks, we study how proximity to these suburban parks benefits the existing firms. We document that these industrial parks offer significant spillover effects: firms outside but close to parks also enjoy productivity growth.[10]

The productivity gains caused by the creation of new industrial parks is not a law of physics, however. In our data set we observe the effects of many parks that differ with respect to which firms enter. We find that there is a larger productivity benefit from being close to a new park whose firms feature a higher level of human capital and a lower number of state-owned enterprises.

The formation of these employment subcenters has implications for alleviating the traffic congestion in the CBD and thus improving quality of life. The growth of such suburban subcenters could also shorten commutes for people who live in the area. It is also possible that people will choose to live even farther from the city center (where land is less expensive) if they work at a suburban location. The environmental impacts of such commutes depends on whether people in China's cities commute to suburban employment centers by car or by public transit. Consider a case where a worker would have commuted by public transit had he worked in the city center but instead commutes by car to a suburban employment center. In this case, the introduction of the suburban employment center increases his carbon footprint because he now drives more.

Rising Demand for Life in the Suburbs

In the CBDs of large Chinese cities, less and less land is available for residential development. This fact has contributed to suburbanization of the urban population. In 2010, more than 70 percent of Beijing's population lived outside the third ring road, and the vast majority of those suburban dwellings are in the form of high-rise buildings. Compared to low-density urban sprawl, this type of high-density expansion is good for the environment because such individuals live in smaller units and have greater access to public transit.

Urbanites' locational preferences change as they age. In the United States, a young single person may prefer the excitement of downtown living, while a married couple with young children might prefer the safety, schools, and open space offered by the suburbs. A couple whose children have moved away from home may be eager to downsize and have a shorter commute with access to culture. In China's cities, young singles and couples without children live close to their jobs so that they can save commuting time and spend more time on their work. When married couples with children choose their residential location, they greatly value proximity to high-quality schools. In China, older people always live with their grown-up children and help take care of their grandchildren.

Wealthy Chinese urbanites will move to the suburbs to enjoy the quiet life and to reduce their exposure to urban pollution. But a problem they face is that the quality of suburban public services, such as schools and hospitals, has lagged behind that of urban areas; wealthier people thus

Figure 4.5 The home page of Jackson Hole's website
Source: http://www.yxmlj.com/yxl/yx.html

continue to be attracted to living in the city center. Some private developers are seizing this opportunity to build primary schools and parks in suburban areas so that more wealthy households will be attracted to their complexes.

We expect that as the middle class grows in China's cities there will be an increasing demand for larger space and more peace and quiet. Mr. Huang is a graduate from Tsinghua University who works in commercial real estate development. Regarding his main residence he greatly cares about access and avoiding traffic, but he would prefer a second, less-expensive home located far from the city. Wealthy people buy single-family houses (villas) as second homes in suburban areas, and they drive there on weekends and holidays to enjoy the quiet life. Basic infrastructure has not kept pace with this private demand.

Siqi's senior colleague, Mr. Li, bought such a villa outside (but adjacent to) Beijing's city boundary. The community is called Jackson Hole, and was developed by a real estate developer who wanted to mimic the developments in Jackson Hole, Wyoming (see fig. 4.5). Journalist Megha Rajagopalan has described life in Beijing's Jackson Hole:

> A two-hour drive north of Beijing's oppressive smog and colorless high-rises, the town provides a surreal sense of escape. Jackson Hole, whose Chinese name literally translates to "Hometown U.S.A.," now consists of about 900 single-family homes with working fireplaces,

wooden facades and landscaped footpaths. Still under construction is a primarily commercial section dubbed Teton Village, which will include a stage for cowboy stunt shows, a gold-panning area for children, and even a church. China's wealthy urbanites have long flocked to tropical resort communities in seaside cities for short-term stays, but ownership of weekend homes in the countryside is a new phenomenon spurred by the explosive growth of the upper class. Many of those homes are purchased purely as investments in a hot housing market, but researchers say country vacation homes make up a growing percentage, though themed resort towns are relatively rare.[11]

Mr. Li told Siqi that he likes to stay in his new home in Jackson Hole when he has free time. The fresh air and the quiet environment provide him with intellectual inspiration, but he also complained about many inconveniences there, including electricity blackouts, winter days without central heating, and lacking easy access to nearby supermarkets. When Mr. Li's family plans to go there on weekends, his wife will buy enough food and daily necessities in the supermarket near their downtown Beijing home ahead of time.

Housing Supply at the Suburban Fringe, and Farm Land Purchases

A distinctive feature of China's cities is the guiding hand of government in determining the allocation of urban land supply, and this affects suburban growth. Given that the inner city has already been developed, there is a greater likelihood of finding available land by seizing it or purchasing it from agricultural communities at a very low price at the city's fringe. The range of compensation for farmers for land taken is often based on income generated in agriculture use instead of being tied to the value of the land if it is allocated for urban use.[12]

To appreciate the income equity implications of this point, suppose that a farmer's piece of land will generate a fifty-year flow of agricultural profits equal to US$100,000 but that a developer could earn US$50 million if he owned the same land. What is the value of this land? While an economist would say that the answer is $50 million, the farmer is being paid closer to $100,000, and the bulk of the gains are going to the local government (through the collected land-sale revenue) and to the developer. This simple example highlights how urban interest groups are using political clout to gain the bulk of the proceeds from land acquisition and land conversion to urban uses. It is important to note that it is efficient for the land to be converted from agriculture to suburban development but that there are enormous equity implications regarding who gains from this asset reallocation depending on how the compensation price is determined. A joint report by the World Bank and the Development Research Center of China's State Council released in 2014 estimates that for all the farm land that was converted to urban land in the past twenty years, the total value gap between what farmers received and the market price is about two trillion yuan.[13]

If cities encounter low cost in acquiring agricultural land, this is an implicit subsidy to suburban growth, and such cities will be more likely to spread out. According to the Chinese National Bureau of Statistics, in the years 1990–2010 China's urban land expanded dramatically, from 12.9 thousand square kilometers to 40.1 thousand.[14] In contrast, if farmers require higher compensation for selling their land, this will incentivize growing cities to permit the construction of taller buildings so that the center city's population density increases.

Farmers whose land is grabbed are dislocated from their old lives and are unlikely to receive enough job training to become urban workers. Recently there have been international media stories about farmers fighting to keep their rural land. The likelihood that such protests erupt depends on the urban-rural land value gap, the level of farmers' education, and whether the local officials have more experience in dealing with such conflicts.

A salient example is the prominent fight in Wukan, which is a suburb of Guangzhou in Guangdong Province.[15] In September 2011, the Wukan village committee sold its collective land to a real estate developer without properly compensating the villagers, who attempted to demonstrate but were repressed by the police. The protest escalated two months later, and these events attracted considerable attention. In response, the Lufeng city government and the Guangdong provincial government sought to satisfy the villagers' requests by giving them a considerable amount in compensation. In many other parts in rural China, villagers do not have the same firm grasp of the nuances of the news media, but Guangdong is China's wealthiest and most liberal province, and it has news publications that are relatively freewheeling. More important, Cantonese speakers there gravitate to the uncensored news reports and cultural products of Hong Kong, on the province's southern edge.

As urban governments become increasingly dependent on land-sale revenue as the main extrabudgetary revenues to pay for urban construction, local politics increasingly revolve around land politics as urban governments seek various means to bring more land under their control.

If farmers enjoy a strengthening of their property rights such that cities can no longer cheaply expand outward, this would provide an incentive for densification closer to the city center. In our research on Beijing we found that the floor area of buildings closer to the city center were not much higher than that of buildings far from the city center. This feature stands in contrast to what is found in US cities. This suggests that the Chinese urban planners could permit the construction of taller buildings closer to the city center and this would allow such cities to continue to grow in terms of population and economic activity without increasing the overall metropolitan area's land footprint.

New Suburban Transportation Infrastructure

Today the Chinese government is investing billions in transportation infrastructure such as highways, a process that began in the early 1990s. Spending on transportation infrastructure grew at 15 percent per year to about US$200 billion in 2007, much of which occurred in cities. Research using US data has documented the role that new highways have played in causing suburbanization in the United States. A chicken-and-egg question arises: Does suburban population growth cause new roads to be built (perhaps due to political influence of the large suburban middle class), or does the construction of new roads lead to suburban growth? In an important contribution to transportation economics, Nathaniel Baum-Snow documents that the United States built roads by following the 1957 Federal Highway Act.[16] This early plan's blueprint is highly valuable to empirical researchers because it allows them to break out of the chicken-and-egg trap; Baum-Snow identifies those highways that were built because they were part of an initial 1957 interstate highway plan. President Eisenhower stressed that this connected highway system would reduce traffic congestion, improve driver safety, and strengthen the nation's defenses. Baum-Snow uses the 1957 plan to estimate the causal impact of a new highway on suburbanization. The US highways were built to connect cities and to increase national security rather than to encourage intercity dispersal. As a result, some metropolitan areas received more interstate highways than others simply because they were located closer to other population centers. Baum-Snow finds that between 1950 and 1990 the construction of new limited-access highways in the United States has contributed markedly to central city population decline.

A group of economists (including Baum-Snow) have been investigating a similar question in the case of China.[17] As China invests in new highways, how do these investments affect the quantity and spatial distribution of economic activities? To answer these questions, the researchers collected information on road maps at several points in time from 1925 to 2005 and satellite data indicating "lights at night" in 1992, 2000, and 2005. Similar to the US research findings, this new China study concludes that highway growth accelerates suburbanization.

As China's metropolitan areas grow there are a variety of environmental impacts including farmland at the fringe being converted into urban space and an increase in greenhouse gas emissions. Rui Wang uses a panel data set of Chinese cities covering the years 2000–2010 and measures a city's compactness by collecting data on its average population density. He finds that as a Chinese city's population density declines (suburbanization) its per capita carbon footprint increases because people are more likely to own a vehicle and to walk less.[18]

Promising Sustainability Trends in Chinese Real Estate

Green Building Demand

In green US cities such as Seattle, many new buildings are constructed to achieve very high environmental performance, such as having zero net energy use and economizing on water consumption. A recent example is the Bullitt Center, which boasts that it incorporates seventeen green features ranging from solar panels to wastewater reuse.[19] Real estate developers will be more likely to build such buildings if the price of electricity and water is high and is expected to rise. In this case, the saved operating costs by "going green" may outweigh the higher up-front costs of erecting such a building relative to a conventional building.

This discussion matters because China's developers are erecting many buildings today. These buildings will last for decades, and there are thus serious implications for aggregate electricity and water demand based on current decisions.

Energy consumption in buildings accounts for 30 percent of the country's total energy use, and this proportion is rising. At an economics forum in November 2011, Tang Kai, the chief planner of the Ministry of Housing and Urban-Rural Development said China erects about two billion square meters' worth of new buildings each year, but less than 100 million square meters are of energy-efficient buildings, which means more than 95 percent of new buildings that go up every year are "energy guzzling." China now has forty billion square meters' worth of buildings, and energy-efficient ones account for a "very small share." China's government knows that promoting the construction of energy-efficient buildings will ease energy shortages and help the country meet its goal of cutting emissions,[20] but existing market institutions (such as the lack of a reliable green rating system and tax-break incentives) hinder the supply of green buildings in the private sector.[21]

In nations such as the Netherlands, certified energy-efficient homes sell for a price premium, and this creates a financial incentive for developers to build such homes. In California, a home with solar panels sells for 3 percent more than comparable homes without solar panels.[22] The degree to which there will be a price premium for green residential real estate hinges on the actual operating expenditure reductions it offers. To appreciate this point, consider an extreme case in which a city's electricity and water is provided for free to its residents; in this case, real estate developers have no incentive to design buildings that economize on the consumption of electricity and water. A key issue with respect to green buildings revolves around who certifies them as actually being green.

In the case of China, special issues arise concerning who is trusted with doing the certification that a building is green and whether green-certified buildings will sell for a higher price. In China, public buildings are easier to certify because the central and local governments have the power to conduct the environmental certifications for them without caring about the extra cost of production. This strategy also allows the government to signal to the public that it cares about the environment. The new building of the School of the Environment at Tsinghua University is a good example of the lingering green building challenge. The building was expensive to erect, and it boasts a world-class design, but the professors and students who work there say that it consumes more energy rather than less, and they are not comfortable inside because it is too hot in winter and too cold in summer. In the United States, advocates of green buildings argue that the buildings both achieve high marks for energy and water efficiency and offer workers and residents higher productivity and quality of life as they offer more sunlight, cleaner indoor air quality, and better temperature control. If Chinese green buildings could achieve a similar level of performance, there would be greater demand for them. This example highlights an important point, that individual users of a building care about their own comfort and productivity. An environmentalist will also care about the greenhouse gas emissions generated by the building. Given that any one building does not cause climate change (because its addition to the stock of global emissions is so tiny), individuals have strong incentives to solely focus on the private benefits of such green buildings. Thus, the key to the diffusion of green building technology is for such buildings to perform well on both private comfort and on energy efficiency. Given that green buildings are a new product in China, many urbanites may know little about the benefits they would gain by living or working in one; this absence of information is likely to lower their willingness to pay to live or work in such a building. To test this hypothesis, Li Zhang (an expert on green buildings), Siqi Zheng, and colleagues conducted an experiment in Beijing to test the role that information plays in promoting home buyers' investment in green housing.[23] They selected two pairs of residential complexes; each pair has two complexes located in the same housing submarket (a small geographic area), and one is a green building while the other is not. Therefore, the location and building quality of the two complexes within a pair are similar except for their "greenness." Their partner, the Building Science Department of Tsinghua University, conducted a field test of indoor environmental quality in December 2014 and designed an information card based on its test results showing that green residences perform much better than their nongreen counterparts in terms of several indoor environmental indicators. Zhang and colleagues used this information card to conduct a before-versus-after information-provision survey in the two pairs of complexes. Before showing this information card to the respondents, they asked each respondent about his or her willingness to buy a green housing unit, and if the respondent said yes, the price premium he or she was willing to pay for it. They then asked the same two questions after showing respondents the information card. The design of the experiment ensures that the willingness to pay change is solely due to the information provided.

The results from the experiment show that those who live in green complexes either have a higher preference for green buildings or already had knowledge about the benefits of living in one. Zhang and colleagues found that green housing dwellers have a higher initial willingness to pay for green residences than dwellers who have not lived in green units; this differential was roughly 40 percent. Additionally, they found that those who lived in the nongreen building increased their average stated willingness to pay for green attributes by more after receiving the information about green technology. This is an optimistic result, because it suggests that education and word-of-mouth learning could accelerate the Chinese urbanite demand to live in greener, more energy-efficient housing.

Such demand for green building is needed to stimulate the market because private real estate developers have been slow to pursue this market niche. In the case of China, private real estate developers are more reluctant to certify their residential towers because of the high costs of certifications compared with the low benefits that green labels have on developers' profits. If residential electricity prices remain at their current low levels (about 0.5 yuan, or eight US cents per kilowatt hour, in Beijing), even a very efficient home offers very little annual benefits to the owner in terms of reduced operating costs. Intuitively, few people would buy a Toyota Prius if gasoline cost nothing per gallon. But according to Ms. Liu, who works for private developer Beijing Vantone Real Estate, the government has recently taken some measures to encourage the development of certified green buildings. "They provide subsidies when Vantone renovates office buildings for platinum certification of LEED [Leadership in Energy and Environmental Design]," she said. "This is attractive for most companies, since 'green' is also a selling strategy." As we will discuss below, the value of green brands is likely to rise in China as the urban public's educational attainment and desire to engage in environmentalism increases.

We interviewed Li Zhang at Tsinghua University. She told us that China launched its own green building certificate in 2008 (one star, two stars, and three stars as the top grade). At first only a very small number of private developers employed this system and got their buildings certified—across the entire nation, four residential complexes in 2008, four in 2009, and forty-five in 2010. But things have been getting better since then. For the years 2011, 2012, and 2013, respectively, 140, 198, and 287 complexes entered this system. But there are still very few certified green buildings given the vast housing market in China. The central government started to offer subsidies for green buildings in 2012. The subsidy is 45 yuan and 80 yuan per square meter for two-star and three-star green buildings, respectively. Cost estimates suggest that this subsidy covers half of the incremental cost for these buildings. Zhang's own research compares those certified green buildings and their comparable counterparts and finds that the former has a 6.4 percent price premium. If this is true, private developers can profit by constructing green buildings.

Besides constructing new green buildings, the Chinese government is also investing in the retrofitting of existing buildings to make them greener. If one measures the total energy consumption per unit, an older building may consume less energy because it is smaller and the households who dwell in such structures often live a simpler lifestyle with fewer modern electrical appliances. But if we compare a new unit and an old one of the same size and the same household lifestyle, the new unit will have a clear advantage in energy saving because it is equipped with modern green technologies. This suggests that the retrofitting of older buildings will generate considerable energy savings.

Green Design and Urban Planning

Chinese urban planners do consider environmental protection in their decision making. Residential and heavy industrial land use are separated to reduce urban residents' exposure to industrial pollution, development is restricted in areas that are ecologically sensitive and habitats for important species, development density is restricted around historical heritage sites, and developers are required to provide green space and open space inside or around new projects. At the typical new residential complex the green space accounts for roughly 30 percent of the complex's total land area. Homeowners pay the maintenance fee for this green space as part of their condominium fee. Other urban planning principals include encouraging transportation-oriented development and introducing bus rapid transit to reduce energy consumption and traffic congestion. These principles are neglected in some cases, however, when they conflict with local governments' pursuit of economic growth. We will return to this point in chapter 9 when we discuss the relative strength of green ministries (such as the Ministry of Environmental Protection) relative to progrowth ministries (such as the National Development and Reform Committee) within the Chinese central government's power structure. This internal power struggle has important implications for a city's ability to change the rules of the game to promote a "green cities" agenda.

Chinese urban planners believe they know "what the best city should look like," but they do not fully understand individuals' preferences and how individuals and firms respond to incentives. Many of these planners did not take neoclassical economics classes when they were in college, but the new generation of urban planners in China has shown great interest in learning urban economics to better understand how individuals' choices aggregate to determine the emerging landscape of a city.

Experimentation with Low-Carbon Eco-Cities

To mitigate the negative effects of growing urbanization and suburbanization, the Chinese central government has invested in the development of China's future cities and has committed to starting to reduce carbon emissions by the year 2030. To achieve this target, energy efficiency and renewable energy are expected to play a major role. The term *eco-city* refers to a wide variety of new urban development models, including the low-carbon and sustainable city model. China's Ministry of Housing and Urban-Rural Development and Ministry of Environmental Protection have jointly created the regulatory framework for the development and construction of approved eco-cities. They recommend performance standards based on economic indicators such as per capita gross domestic product (GDP) and energy and water consumption per unit of GDP; environmental indicators such as air and water quality, solid waste treatment rates, and per capita urban green space; and social indicators such as urbanization level, Gini ratio, and public satisfaction with the environment. There are currently more than 150 eco-city projects at various stages of development in China.

Two examples of eco-city projects are the Sino-Singapore Tianjian Eco-City and the Baoding low-carbon city. The Sino-Singapore Tianjian Eco-City (SSTEC) project was launched in 2007 from a collaboration between the government of Singapore and the Tianjin municipal government.[24] Located forty kilometers east of Tianjin (the city with the fourth largest urban population in China) at the perimeter of the Binhai New Area, this new development covers 34.2 square kilometers. The full completion of the project is targeted for 2020, and is expected to be home for 350,000 residents.[25]

The eco-city seeks to be green by introducing some of China's strictest energy efficiency standards, and green power is produced via both wind turbines and solar panels, but the *New York Times* reported in summer 2015 that the wind power had not yet been connected to the grid. Another metric for being a green city is walking, biking, and the use of public transit rather than cars for transportation purposes. A green city is more than its built infrastructure, and ideally such a city both selects people who want to live the green lifestyle and has a causal effect on shrinking a person's pollution impact. According to the *New York Times* article, "Critics also say that little is being done to educate residents about energy efficiency and recycling. Although neighborhood centers intended to teach people about sustainable practices will be scattered across the eco-city, each would serve an average of about 30,000 to 35,000 residents. The eco-city has a more complicated recycling system than the rest of the country, with residents being asked to separate their discards into five categories."[26] This quote highlights an interesting question: Will China's investments in green places increase environmentalism among its people? In work based in the United States, Matthew E. Kahn has documented that people with a higher education level are more likely to engage in environmentalism.[27] It remains an open question whether living in an eco-city also causes a rise in embracing a "green lifestyle."

The main innovation of this eco-city project is its layout. To minimize the need for commuting, SSTEC consists of integrated mixed-use zones in an "ecocell" layout, a modular sixteen-hundred-square-meter grid that is repeated throughout the site. Four to five ecocells are combined to form "eco-neighborhoods" that can accommodate twenty thousand residents with mixed housing types to avoid the formation of ghettos.[28] The mixed land-use plan will integrate housing with service-oriented, high technology, and environment-related industries that are anticipated to create 190,000 jobs, and the project incorporates a comprehensive green transportation network based on nonmotorized and public transportation. The majority of daily services and basic necessities will be available to residents within a five-hundred-meter radius.[29] When it comes to resource management, water conservation techniques such as recycling of domestic and industrial wastewater and rainwater harvesting will be employed, so much so that the use of nontraditional water sources is expected to reach 50 percent. Through energy conservation and energy-saving manufacturing processes, per capita energy consumption will be at least 20 percent lower than the national average. By May 2010, the SSTEC project had attracted over seventeen billion yuan of funding, 125 companies have already registered for a combined capital of over ten billion yuan.[30]

Whether such low-carbon eco-cities represent a cost-effective strategy for promoting the rise of a more sustainable urban China remains an open question. The key point is that China's cities are experimenting. Each of these local government initiatives can be thought of as field experiments that will generate new knowledge about which policies work and which do not. When a good idea is identified, such an initiative can be scaled up and repeated in other locations. This example highlights that good ideas are public goods, and the key is for local governments to be honest that they "know that they do not know" which is the right path to pursue. Through experimentation with many strategies, the most cost-effective green policies will be identified.

The likelihood that innovative policies that have been demonstrated to be effective in one location are repeated in other areas depends on the political clout of key environmental agencies within local and national government. Some scholars are cynical about the eco-city experiments. Some earlier adopters of these new ideas were promoted to high positions within the government, which incentivizes local governments to compete for titles such as green city, recyclable economy city, and low-carbon city. Local officials are more or less interested in green *planning* rather than doing something concrete. Some researchers have argued that one factor slowing down the diffusion of effective green policies is fragmented and weak environmental bureaucracies.[31]

The rise and fall of the Baoding low-carbon city, a municipality in Hebei Province, provides a cautionary example. The project is over ten years old, and its low-carbon initiatives were initially projects carried out under the High-Tech Administrative Committee, a small, low-key agency that has pushed this agenda since 2001. The agency faced strong opposition from the Baoding city government and other departments. Clean energy technology firms—those involved in various solar photovoltaic (PV) and light-emitting diode (LED) projects—started to flourish in the high-tech zone. Only in 2006, after some early successes and national policy shifts, did the Baoding city government become interested and begin to support the low-carbon city. But the project has been stalled since 2010, when it was transferred to the more powerful city government (the Development and Reform Committee and the Bureau of Science and Technology), which has been noncommittal about further carrying out the low-carbon agenda. For example, few residential areas have been retrofitted with solar PV panels and LED lamps since 2010, and some earlier renovation projects have in fact again been replaced with traditional technologies. The city did not have the human capital and the skill to maintain the solar PV and LED projects, and because they did not offer a consistent stream of services, the city blamed the immaturity of these technologies.[32]

This pessimistic example highlights that the prospects for low-carbon cities to develop hinges on the cost of electricity and water (i.e., the saved operating costs by going green) and on the scale of the green economy. If China's central government commits to developing specific green technology, young workers will invest in human capital that allows them to specialize in green jobs. Their collective investments in these new skills would mean that cities such as Baoding could confidently deploy green technologies without fears of intermittent service. The example suggests that if weak government agencies are the prime proponents of pursuing the green economy, this may not be a large enough "green push" to encourage sufficient investment of financial or human capital to give the green economy a chance to succeed. The good news (as we discuss in chapter 9) is that the central government is making increased investments in pursuing a sustainability agenda.

This chapter has focused on the trade-offs of living in the center city versus the suburbs of a big city. In the United States, suburban life often revolves around using one's car to travel from one place to another. Chapter 5 explores the rise of car use in urban China.

CHAPTER FIVE

Private Vehicle Demand in Urban China

Today China is the world's leading producer of greenhouse gas emissions, but not too long ago its carbon transportation footprint was quite small. This was due to its poverty, but also to the way that society was organized under Mao Zedong. Mao's original urban planning vision actually minimized transportation costs and achieved a low carbon footprint from transportation. During the planned economy era, land resources were allocated according to central planning; each work unit constructed residential buildings on its own land. Employees lived in the work unit's area and usually only needed to walk or ride a bicycle to commute to work.

Mao's plan minimized transportation externality costs (there were no cars at that time) but ignored the agglomeration productivity benefits of different types of firms locating close to each other. Jane Jacobs, the famous urbanist, argued that much of New York City's vibrancy was due to the learning that takes place when different types of firms are in close physical proximity. In Mao's China, firms engaged in productive activities according to centralized planning, and these self-contained firms gained little from locating close to each other. In recent years, as China's cities have embraced market forces, urban economic activity has reorganized to maximize the productive agglomeration forces of firms that can trade with and learn from each other.

As Chinese households have grown wealthier, their demand for private automobile use has soared. Such vehicle use saves households time and offers a degree of comfort and reduced exposure to air pollution on smoggy days relative to commuting by bicycle or using public transit.

In 1990 there were only 12.8 million motor vehicles in China, but by the year 2008 this count had grown to 49.6 million; by the year 2020, there could be two hundred million vehicles in the nation. China's total population is roughly 1.3 billion people. If it achieves the US vehicle ownership rate, there will be slightly more than one billion vehicles driving around China.

In this chapter we discuss why more and more Chinese urbanites demand cars and examine what environmental challenges are exacerbated by their use. Our starting point is that speed in cities is essential. Given that time is our scarcest asset, increases in urban speed have played a central role in improving urban quality of life and raising the likelihood that urbanites have a high standard of living. The ability to move faster in cities makes urban living more valuable; it allows people to interact with more friends, employers, and potential customers and to shop at more stores. The ability to travel to farther job locations and to business meetings means that workers are more likely to be matched with the right employer, and this enhances their productivity, which in turn translates into greater macroeconomic growth.

Hundreds of years ago, when walking and horses were the only transportation modes, urbanites could only trade and interact with a limited set of people and firms. One could not live more than ten miles from where one worked. Transportation innovations ranging from streetcars to subways and private vehicles have increased the trading and learning possibilities that cities offer. If a person can now move at a speed of thirty miles per hour in a city, in thirty minutes he or she can reach all locations within a fifteen-mile radius.

As China's urbanites grow wealthier over time, they seek to save time commuting, and private vehicles are often the most effective transportation mode for saving time by moving at the highest speed. It is no accident that all over the world, wealthier urbanites have sought to own and drive cars. In this chapter we seek to explain current patterns across China's cities and make some informed predictions about future driving patterns.

Transportation Facts

In 1950, there were two thousand vehicles in Beijing. By 2011 this number had risen to 4.98 million, of which 3.9 million were private cars. In 2020, there may be as many as ten million vehicles in Beijing. Fast-growing car ownership is taking place in many large Chinese cities. A large-scale survey conducted by the National Bureau of Statistics of China in 2010 reveals that, in all cities at the prefecture level or higher, 11.2 percent of households own private cars, yielding a ratio of thirty-seven vehicles per thousand people. In 2009, there were over 254 million registered vehicles in the United States;[1] this works out to roughly .8 vehicles per person (including children and seniors!). This chapter will investigate the causes and environmental consequences of the likely convergence in ownership rates that will take place between the United States and China over the next few decades.

The distribution of vehicles across Chinese cities is uneven. Based on a large-scale survey, in wealthy cities such as Beijing, Hangzhou, and Shenzhen, more than 30 percent of the households own private cars, while in some small- and medium-size cities located in China's underdeveloped areas (such as Datong and Lanzhou) this share drops to less than 5 percent.

The Vehicle Purchase Decision

The simple economics of purchasing a vehicle states that a potential buyer will compare the benefits and costs of such an investment. Cars are costly to buy and operate. Herein we will sketch their costs in China today. The benefits of car ownership hinge on how an urbanite has configured his or her life. Even people who can afford a vehicle will not do so if they live in a dense city with high-quality public transit.

In wealthy Manhattan, relatively few people own cars. Coauthor Matthew E. Kahn's parents sold their car soon after they moved from suburban Scarsdale, New York, to Manhattan because they were paying six hundred dollars per month to park it and were rarely using it. Manhattan is known as a place of pedestrians and public transit; its population density of 67,000 people per square mile is the highest in the United States. Given the large fixed costs to building subway infrastructure and the economies of scale in bus service, areas with higher population density are likely to have better public transit infrastructure, and this will decrease private vehicle use. The 2004 National Economic Census figures for Beijing show that there are 7,000 persons per square kilometer and the employment density is around 14,700 persons per square kilometer in the four inner-city districts. By comparison, Tokyo's four central districts have an average employment density of 47,000 persons per square kilometer; Beijing is actually not very dense because of the low-density buildings built before the 1980s in the central city and the enforcement of urban planning principals discussed in chapter 4.

Unlike Matt's parents in Manhattan, coauthor Siqi Zheng has owned a car for seven years now, and her situation is typical of successful people in Beijing. Siqi moved to the city in 1995 when she entered Tsinghua University as an undergraduate student. She got married in 2003 when she was twenty-six years old, and started her career as an assistant professor at Tsinghua in 2005. She purchased her first car in 2007 when she was thirty years old. Her new Dongfeng Peugeot car cost her 120,000 yuan (roughly US$18,000).

Her parents do not own a car, and they do not know how to drive, so Siqi drives her parents to dine at local restaurants. Since driving private cars was not that common in Chinese cities prior to 1990s, many of the older generation did not learn to drive when they were young and now they are reluctant to learn; in addition, most young people do not have the money to purchase cars. This means that middle-age people are the most likely to own and drive cars.

Many couples share the same story: they bought their first car right after they started to earn a salary. When such couples have a new baby, a car becomes a necessity because they do not want to travel with their child on a bicycle or on the congested buses and subways. Ms. Wang is Siqi's good friend. She also bought her first car two years after she received her doctorate from Tsinghua, and two years after that she had a baby girl. Most of Siqi's high school friends back in her hometown of Zhengzhou (a medium-size city in China's central region), who are in their thirties, have bought cars.

Young Chinese couples prefer to buy new cars instead of used ones. While in many developing countries imports of used cars represent a large share of vehicles, in China most purchases are of new ones. The price of domestic-produced vehicles in China is not that high (a new medium-priced car costs around 100,000–150,000 yuan ($US16,000–24,000), but there are large tariffs applied to new cars imported from other countries (mainly from Japan and Europe); a new import costs around 750,000 yuan. In China many car manufacturing companies are joint ventures with international concerns; Guangqi Honda, Shanghai General Motors, Shanghai-Volkswagen, Yiqi Honda, and Yiqi-Volkswagen are some of the most popular brands. Buyers can pay cash or finance using car loans. Siqi's and Ms. Wang's new cars were not expensive, so both of them paid using cash. Some of their friends who are saving for a new condominium chose to finance their car purchase through a loan.

There are hundreds of Chinese car brands, with various characteristics to choose from, so there is quite a competitive new-car market. The quality adjusted price of Chinese vehicles has been falling sharply over time at a rate of roughly 8 percent per year.[2] A household could purchase the same vehicle in the year 2019 for half the price it paid in 2014! Consumers benefit from declining car prices, and this means that more middle-class people will purchase their first vehicle.

Prices are declining because of the fierce competition among Chinese carmakers and industrial learning by doing such that domestic makers are becoming more productive. In 1980, China's domestic vehicle industry produced five thousand passenger vehicles; by 2011 that number had risen to fourteen million. Rising incomes and falling real prices have made China the world's largest auto market since 2009. In recent years, due to such joint ventures such as Dongfeng collaborating with Honda, Beijing Automotive Investment with Hyundai, and Shanghai Automotive collaborating with General Motors, prices for vehicles have declined as their quality has improved. There are four possible explanations for declining vehicle production costs: (1) industry-level learning by doing that lowers cost of production for all domestic firms; (2) tariff reduction due to China's joining the World Trade Organization; (3) the appreciation of the yuan relative to the US dollar, which lowered the cost to China of iron, steel, and aluminum;[3] and (4) the reduction in the cost of auto components due to the growth of domestic supply chains, specialization, and competition in China. Such supply shifts help explain how domestic vehicle prices can be falling at a time when demand is rising.

What Do Car Buyers Want?

Chinese urbanites who own cars use their vehicles chiefly for two types of trips: commuting and weekend travel. An attractive feature of private driving is its comfort: wealthy people and the middle class flee crowded public transportation and enjoy privacy, air-conditioning, and music in their own cars. "Although the traffic is bad, I would rather spend my commuting time in a comfortable space than taking the subway," notes Ms. Lin, one of our interviewees.

Chinese car buyers trade off vehicle attributes such as size, engine power, fuel economy, and the carmaker's reputation. Ms. Chen, who received an MBA in the United States a few years ago, works as an executive in an investment company. She told us that her priorities have evolved over time in deciding which car to purchase. "When I bought my first car, I considered the car's price as the main factor," she said. "But now, I pay more attention to brand, since gasoline prices are no longer a problem for me." While the very wealthy buy expensive cars with powerful engines as a symbol of their social status, most Chinese car buyers do not prefer large cars; as gasoline prices have increased, the sales of larger cars has declined in recent years. (As economists, and as environmentalists, we are pleased to see evidence that Chinese demand curves slope down.) In the United States, researchers have documented that a consequence of the Organization of the Petroleum Exporting Countries' oil shocks of the 1970s was the increasing demand for smaller, more fuel-efficient vehicles; Japan's export market boomed as companies like Toyota supplied such vehicles. In 2015, as global gasoline prices decline, it will be interesting to see if Chinese consumers trade up to larger, less fuel-efficient vehicles.

There are considerable price differences for cars with different engine sizes. For example, an average domestic car with a 1.6-liter engine costs about 120,000 yuan, but a car with 2.4-liter engine costs about 270,000 yuan. When Siqi bought her new car in 2007, her priorities were safety and roominess (she wanted her parents, husband, baby, and herself to all sit comfortably in the car), and she wanted the car's trunk to be large enough to hold a baby carriage. She also cared about its fuel economy (she did not want to pay a lot for gas because she was at the start of her career). Siqi and her husband shopped and negotiated with five car dealers before she finally chose her 1.6-liter Dongfeng Peugeot.

We now further investigate other key costs of owning and operating a vehicle; these include paying for gasoline, annual automobile insurance, parking, and maintenance.

Gasoline Prices

Today most cars in China are gasoline powered. Before 2005, gas prices in China were much lower than that in the West, so driving was subsidized, but in recent years gas prices have been rising. Today, the price of gasoline in China (roughly five US dollars per gallon) is roughly 30 percent higher than in the United States, but is still lower than in France, Japan, South Korea, and the United Kingdom. The price is controlled by the National Development and Reform Commission (NDRC). Unlike other sectors of the economy, gasoline is supplied by three big state-owned enterprises (SOEs): China National Offshore Oil Corporation, PetroChina, and Sinopec. The official position of the NDRC is that the purpose of their central pricing policy is to prevent these three big SOEs from monopolizing gas pricing and to ensure that domestic pricing keeps pace with international price dynamics. Some have claimed, however, that these three SOEs are so politically powerful that the NDRC takes orders from them and meets their demand for increasing gas prices. Such "regulatory capture" is a major theme in US regulatory studies. From a purely environmental point of view, it may even seem good for air quality if the three oil SOEs encourage the NDRC to raise gas prices, but they also put pressure on the NDRC to delay the implementation of tighter gasoline refining standards (as revealed by the recently popular documentary *Under the Dome*, produced by Jing Chai), and it goes without saying that this is bad for the environment. When powerful firms are able to dictate policies that enhance their own profits, the public interest can suffer.

Insurance

Chinese drivers must have insurance for their vehicles. Throughout the world, insurance is priced as an annual fixed cost, but in an ingenious proposal, Aaron Edlin has argued that it should be tied to the number of miles one drives in a year.[4] Consider two identical drivers. The one who drives less would be charged a lower insurance rate; such marginal pricing would reduce pollution, traffic congestion, and auto fatalities, as individuals would have an incentive to economize on their driving. Despite the wisdom of this idea, we know of no nation that has adopted this pricing scheme. In China the annual insurance fee is about 3–6 percent of the total purchase price of a car, and the insurance premium varies according to a car's age. It grows gradually over time after the vehicle purchase, as the car depreciates; older vehicles have a higher probability of breaking down. Insurance prices do not vary by the driver's gender or age. The insurance company does increase one's premium if a driver is involved in an accident and claimed compensation from the insurance company in the previous year. Back in 2011, Siqi was not lucky; once she drove her car into a tree while parking, and another time when driving slowly in congested traffic the car behind her bumped her car. The result was that her insurance fees increased by about four hundred yuan (10 percent) for 2012.

Parking

The car owner must pay for parking at the origin and the destination. In China, more than 80 percent of urban households live in middle-rise and high-rise condos. Less than 10 percent live in single-family houses. Those condominium complexes built before 2000 have very limited parking space, because at that time urban planners and developers did not expect so many households to own cars. Some complexes do not even have separate parking spaces (either underground or on the surface), so residents have to park their cars in every available public space, resulting in very crowded streets (see fig. 5.1). One professor who works at Renmin University told us that he does not like to drive his car because he worries that there will be no place to park when he returns home.

Residential complexes completed after the year 2000 have, however, built sufficient parking space. In Beijing the building code requires that at least one housing unit should be matched with one parking spot. Those parking spots are sold or leased to the residents who live in the complex. Siqi's family lives in a residential complex completed in 2005 that is close to the Tsinghua campus. Before moving there, she rented a housing unit in an old complex with no formal parking space at all; it was rather congested, with cars parking everywhere at night, sometimes even blocking the fire lane. Although the situation has been greatly improved in newly built residential complexes, the lack of parking space is still a major problem because many households now have more than one car. Siqi's family

Figure 5.1 Cars parked in the public spaces of housing complexes

rents a parking spot in their complex at a monthly cost of three hundred yuan; the price has not changed since 2005 because people will protest if there is an increase. Siqi's good friend, Ms. Ren, and her family moved to their new house in a newly completed complex in 2010, and the monthly rental of a parking space costs nine hundred yuan! Everyone now knows that parking spaces are valuable, and the recent market value of a parking spot has been rising sharply as people anticipate that more people will own cars or buy second cars. Investors recognize that owning parking spots is a wise real estate investment.

University of California–Los Angeles professor Donald Shoup has earned national recognition for his persistent emphasis of the high cost of "free parking" in major US cities.[5] He has argued that the misallocation of parking in cities such as Los Angeles contributes to traffic congestion as drivers cruise slowly looking for a free spot. If parking was priced to reflect its true social cost, less urban land would need to be paved for use as parking lots. Los Angeles zoning laws require that real estate developers provide free parking that is bundled to the ownership of a condominium.

Urban residents must purchase this bundle even when if do not want the spot. When Matt moved to Los Angeles in 2006, his family rented an apartment in Westwood Village; his rent covered both the apartment and two parking spaces.

Shoup has sought to design incentives to reduce the amount of parking in cities and to save urban residents time. He argues in favor of time-of-day pricing for parking such that at peak times meters charge a higher price per hour and prices fall when demand is low. In the United States, cities like San Francisco are embracing his ideas and are experimenting with parking rates that vary over the course of the day.[6] The widespread use of smartphones and the development of applications for them lowers the implementation costs of cities introducing time-of-day parking prices. Without distracting the driver, a set of parking meters could send a signal of their current price and the application would speak so that the driver could hear the stated price; the price would adjust so that demand equals supply. The on-the-ground reality would be that people could always find a place to park and would thus "cruise" less for free parking. The net result would be less traffic congestion, less pollution, and higher road speeds.

Parking costs are higher during the daytime (7:00 a.m.–9:00 p.m.); for example, in the Beijing city center, fifteen minutes of parking time costs 2.5 yuan for a small car in the daytime, but only 1 yuan for an hour at night. In addition, parking prices are higher downtown than in the suburbs. This adoption of dynamic pricing suggests that the Communist Party appreciates the insights from free-market economics concerning how to allocate scarce resources—in this case, urban land.

When going to work, drivers must pay for parking on the streets in the central business district (CBD) and in office building parking areas; the parking fee is regulated by the municipal government's Development and Reform Committee. Before 2010, the parking fee in Beijing was quite low—in most places an hour parking only cost two yuan (less than thirty US cents), but later the municipal government realized that this low cost contributed to congestion in busy areas; it then increased the fee in the innercity. The highest parking rate in the CBD is fifteen yuan per hour. Generally speaking, parking prices are gradually increasing in big cities as local governments implement higher vehicle taxes and enforce stricter parking laws.[7]

Some big companies provide their employers with free parking, but many small companies (especially those located in the CBD) are unable to do so. Universities always have large campuses, so they provide their faculty members with free parking; Siqi parks her car just below her office building. A new problem on campus is that as more graduate students have cars, the demand for parking exceeds supply. This trend will soon force the university to charge students and professors for parking spaces. (Communist China is using market principals to allocate scarce resources efficiently!) The net effect of such higher parking fees is to discourage car trips on the part of individuals who were at the margin of riding a bike or carpooling.

The total cost of vehicle operation can be easily seen with an example. If the typical household drives five thousand miles each year and owns a vehicle that achieves thirty miles per gallon, this household will require 167 gallons of gasoline and will pay US$833 for it. This household also needs to pay $625 in auto insurance per year. Adding regular maintenance expense, the total cost of driving is about $1,500 per year. Given current Chinese urban incomes, this is a fairly large budget share. Some senior managers in private companies and public institutions can have their gas bills reimbursed by their employers; it becomes part of the companies' operating costs. As a result, these individuals respond to this implicit compensation scheme by being more likely to own a car and to drive it more.

Alternative Modes of Transportation

Relative to driving in one's own car, public buses are slow and less comfortable. When you ride the bus, you must walk to a bus stop, wait for the bus, and then sit as the bus stops frequently to pick up and discharge passengers; the net result is time loss. In a growing economy where "time is money," wealthier people seek out the more expensive but timesaving commuting option, the car, while poorer people take the bus. Public buses are more likely to attract wealthier riders who use smartphones if the bus provides free Wi-Fi and if new applications provide real-time information about when the next bus will arrive at a specific location. Some cities have adopted bus rapid transit so that buses have dedicated lanes to achieve higher speeds. Such improvements in quality could allow this green transportation mode to gain a higher market share.

Chinese subways differ from cars and buses in that they are fast and inexpensive to ride; the nation has made an enormous investment in new subways. In 1980 Beijing was the only Chinese city with a subway, but by 2000, three Chinese cities had subways. By 2010, twelve cities had subways and light railways, and as of 2015 at least sixteen other cities have lines under construction. As of 2011, Beijing had 372 kilometers of subway lines and 218 stations; Shanghai had 413 kilometers of lines and 278 stations. Daily passenger ridership exceeds 7 million and 5.5 million in these two superstar cities, respectively. According to a large-scale survey conducted by the National Bureau of Statistics of China for 2010, the shares of commuting by subway among total commutes in Beijing, Shanghai, and Guangzhou were 11.8 percent, 8.2 percent, and 5.9 percent, respectively.[8]

Subways tend to make center cities stronger as they connect people to downtowns with a fast, inexpensive mobility option. Unlike many US cities, China's cities feature the population density to make subways a viable urban commuting option. Beijing has made enormous investments in subway infrastructure; five new subway lines were built between 2000 and 2009 at a cost of $50.3 billion yuan. Figure 5.2 displays the locations of the five new subway lines.

The routes were not chosen at random. When deciding where to build the subway lines, the government considered several factors. The first was to mitigate current road congestion or to meet the anticipated ridership growth (especially for the subway stops in and around the city center). For instance, the Beijing Municipal Commission of Urban Planning recently declared that subway lines 6 and line 7 would be constructed to cope with the ridership growth of subway line 1 and the road congestion around the Beijing West Railway Station. The Beijing municipal government regarded subway construction as a basic infrastructure provision intended to encourage growth to the previously underdeveloped areas. The history of urban development in Beijing left an important urban legacy—North Beijing, where most of the government branches, universities and schools are located, is more developed and wealthier than South Beijing. The Beijing municipal government aims to promote the development in South Beijing by investing in more infrastructure projects there. New subway lines also extend to surrounding satellite towns to support fringe development.

In Chinese cities, bus and subway fares are heavily subsidized. In Beijing, buses have a universal fare of 0.4 yuan per person no matter how long

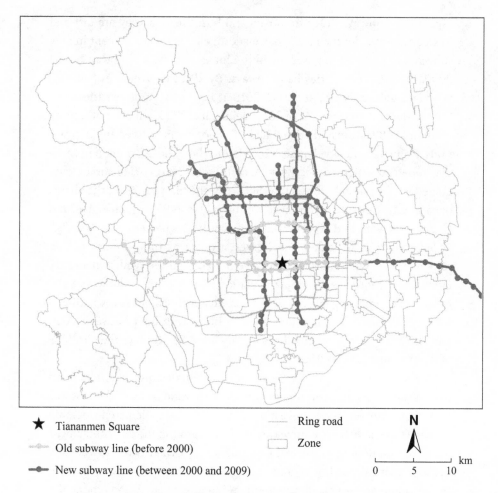

★   Tiananmen Square

Old subway line (before 2000)

New subway line (between 2000 and 2009)

Ring road

Zone

N

0   5   10   km

Figure 5.2 Beijing's recent subway lines

one travels. The universal fare price for subways is 2 yuan per person (about thirty-five US cents) without a travel distance restriction. This low pricing policy is part of the Public Transit Priority policy, which imposes a heavy fiscal burden on the Beijing municipal go vernment. In 2010 Beijing subsidized bus and subway operations with 13.5 billion yuan, accounting for 5 percent of the city's total fiscal expenditure that year.

Trucks and Taxis in Cities

Within cities, bustling stores and restaurants must restock inventory each morning. Construction sites need cement, concrete, and steel from manufacturing firms outside the cities. As such, the trucking network plays a crucial role in serving such logistical. To conserve on gasoline, many trucks take diesel fuel, but these trucks tend to be high emitters of particulate matter.

The megacities in China tend to be major international business centers, and this creates profitable opportunities for thousands of taxis. It is easy to catch a cab in Beijing during nonpeak hours; you can always get one within five or ten minutes. But on bad-weather days and during peak hours, you may have to fight with other passengers for a passing empty cab. Taxi fares are controlled by municipal governments; in Beijing you need to pay ten yuan for the first three kilometers, and each additional kilometer costs two yuan. The idling time when stuck in traffic also costs money—five minutes' idling time costs one yuan. Siqi pays about fifty yuan to get from the Tsinghua campus to Tiananmen Square by taxi if there is no congestion. But this amount will increase to eighty yuan (eleven US dollars) on a busy day.

Taxis are owned by hundreds of taxi companies in Beijing, and taxi drivers have to pay a fixed lease fee to their companies. Those drivers earn the difference between their gross revenue minus the lease fee and the cost for gasoline and regular vehicle maintenance. This residual profit is shrinking due to rising gas prices as well as rising lease fees. Some taxi drivers left the industry when they found that their income could not even support their daily life. Many of Siqi's friends who do not drive use taxis.

In the United States, many people now use Uber, a car service application, as a substitute for taking a taxi. Uber offers convenient pickups, offers higher-quality vehicles than the standard cab, and payment is prearranged so there is no need for cash on hand. In Beijing there is an Uber-like car service called Yongche. Siqi installed its application on her iPhone, and has linked her credit card to the app. This service is more convenient and comfortable than taxi, but it is 50–100 percent more expensive than a taxi (though Yongche offers an airport discount). In the last six months, two other companies, Didi and Kuaidi, have entered the business and now there is fierce competition among them. An increasing number of people have started to use such services.

The net environmental effect of this mode shift in travel hinges on whether the new riders on Yongche used to travel via public transit or their own vehicles. If this higher-quality service is attracting past public transit users, its growth unintentionally contributes to local and global environmental pollution.

The Rising Traffic Congestion Challenge

The spatial separation of jobs and housing leads to a high demand for transportation. With a limited supply of transportation infrastructure, this will cause congestion, which has become a pervasive phenomenon in many large Chinese cities. For example, the average one-way commuting time in 2006 Beijing was thirty-eight minutes.[9]

The people of Beijing wasted seven more minutes stuck in traffic each day in 2009 than they did in 2007.[10] Given that the roads are public property, a type of tragedy of the commons emerges: no one driver has an incentive to recognize that his driving at rush hour slows down everyone else. People could drive at faster speeds and thus save time if there was time-of-day pricing on the roads, such as London's central congestion charge. In London, drivers who enter the city's center during peak hours pay a hefty fee for doing so.[11]

Until now, China's cities have not introduced such pricing incentives, but discussions among policy makers and scholars are taking place. Economists disagree with traffic engineers who argue that new road expansions can mitigate the problem of road congestion. In the absence of charging a pricing for driving at peak times, the fundamental law of traffic congestion predicts that car owners will take more discretionary trips (i.e., drive to a Starbucks coffee shop more often) whenever they believe that the roads are uncongested. The net effect of this dynamic is that new road construction simply leads to more trips, not faster commute times. Urban economists Gilles Duranton and Matthew A. Turner recently studied US transportation investments and resulting road speeds from 1970 to 2010 and see sharp evidence supporting the law of traffic congestion. Their findings indicate that merely increasing the supply of road lanes will not solve congestion because new car trips will result and the new lanes will be quickly clogged again.[12]

To mitigate traffic congestion, major Chinese cities have implemented the Public Transit Priority policy. Dedicated bus lanes are assigned on major roads, and cars cannot enter these lanes during peak hours. In such cases, buses can move at higher speeds than cars, picking up passengers only at designated areas. Another example is the Bus Rapid Transit (BRT) system in Guangzhou. Its first line was put into operation on February 10, 2010, and the system handles approximately one million passenger trips daily with a peak passenger flow of 26,900 passengers per hour per direction (ranking second in the whole world after the TransMilenio BRT system in Bogota, Colombia). In fact, this rapid transit system contains the world's longest BRT stations—around 260 meters, including bridges—with bus volumes of one bus every ten seconds or 350 per hour in a single direction.[13] In 2012 the Guangzhou BRT was named by the United Nations Framework Convention on Climate Change as one of nine "lighthouse activities" in developing countries. Such activity either helps to curb greenhouse gas emissions or help people adapt to climate change, while at the same time benefits the urban poor.[14] Some people say that Guangzhou's network of BRT lines will serve as a model for Chinese cities struggling with traffic congestion.[15]

Local Air Pollution from Transportation

Chinese cities have high levels of particulate matter. In 2013, the annual average particulate matter concentration (measured as $PM_{2.5}$) exceeded the standard of thirty-five micrograms per cubic meter in more than 90 percent of the seventy-four cities for which data are available. In Beijing, motor vehicles emissions produced 33 percent of the inhalable particles, 64 percent of the oxynitrides, and 51 percent of the hydrocarbons in Beijing's air. Motor vehicle exhaust gas has become Beijing's largest source of air pollution.

Siqi's home is a twenty-minute bicycle ride from the Tsinghua University campus. During peak hours she spends even more time on road if she chooses to drive. On good weather days she prefers to ride her bicycle, but a negative result of riding her bicycle in busy traffic is that she has to breathe the dirty auto exhaust.

Diesel-powered trucks contribute a large share of local air pollution in Beijing. A recent regulation in Beijing states that trucks cannot enter the metropolitan area (inside the fifth ring road) during the daytime (7:00 a.m.–8:00 p.m.). But Beijing cannot refuse those trucks at night—goods for sale and for construction have to be shipped in from outside the city. There are thousands of construction sites in Beijing. The good news is that since the emission standard is being upgraded over time, many old trucks have been phased out in large cities such as Beijing and Shanghai.

In many major cities, environmentalists have lobbied for truckers to swap out their old trucks for cleaner trucks with new technology. These new trucks are quite expensive and the truckers have strongly opposed the change. At the end of the day, this is a property rights issue: Do truckers have the right to use whatever truck they deem best for delivering goods, or can society mandate the technology used to deliver goods? Given that there is no free lunch, if cities such as Beijing forbid old trucks to be used, end consumers will face higher prices for goods. Most of China's growing bus fleet uses gasoline. Some use liquefied petroleum gas or compressed natural gas (CNG), fuels that are greener. In recent years, US cities with high levels of air pollution have phased out all of their diesel buses and substituted hybrid and CNG buses.[16]

China still has relatively lax emissions standards for heavy vehicles (in particular, dump trucks), which generate large amounts of dust as they travel. Though some major cities have implemented strict restrictions for heavy vehicles in the daytime, they still need such vehicles for construction projects and industrial production, so local governments allow some to drive at night.

Public Transportation Policies That Will Lower Urban Pollution Levels

The previous sections have emphasized the microeconomic factors that determine the scale of transportation. In aggregate, such individual choices have important implications for total gasoline consumption. In this section we discuss a number of public policies intended to reduce the social cost of urban transportation.

Emissions Taxes and Congestion Fees

Most economists favor direct pollution taxes in order to incentivize the producers of pollution to internalize the social costs of their actions. One applied study concluded that the optimal gasoline tax for the United States would be one dollar per gallon higher than the current tax.[17] A recent study using the same methodology but focusing on China concludes that China's current gasoline tax of one yuan per liter (roughly fifty US cents per gallon) is too low a tax to encourage Chinese drivers to internalize the social costs of local pollution, greenhouse gas production, and traffic congestion associated with their choice to drive.[18]

Improvements in information technology have made it easier for local authorities to introduce an emissions tax for driving. Using odometer readings, the authorities could calculate the total miles driven during the previous year. This mileage could be multiplied by an emissions factor based on an annual smog test that measures the vehicle's flow of emissions. Such a pollution tax would incentivize a car buyer to own a cleaner, newer car and to drive it less.

This incentive regime has not been adopted, and the tragedy of the commons continues: each driver has little incentive to internalize the social costs associated with the small amount of emissions each car contributes to the whole. There has been a long policy debate on the feasibility of this pollution tax in China, but the government has not adopted one yet. Besides the technical challenge of accurately measuring a car's externality and translating it into monetary terms, one possible reason is that China wants to stimulate its car manufacturing industry and is thus reluctant to implement the tax because of a concern that it would reduce new-car demand.

From the individual consumer's point of view, a pollution tax would raise the short-run operating cost of driving a vehicle; it would pose an equity burden on lower-middle-income groups, who would pay a larger share of their disposable income on operating their vehicles. Those in this group are likely to only be able to afford secondhand cars or inexpensive cars, both of which generally cause greater pollution.

In Shanghai the local government is considering implementing a road pricing policy in the Inner Ring Expressway, but there are a number of details that must be hammered out, including how to define the geography of the congestion zone and what price to charge.[19] Such a pricing system would have a different impact on different socioeconomic groups. The upper class will not be discouraged from using their cars, and reduced congestion may give them an incentive to live farther from their jobs. On the other hand, the lower middle class may choose to take public transportation. The same issues arise in the United States. Wealthy people would prefer to pay a monetary fee to access fast roads because it will save them valuable time; poorer people prefer to pay no fee for accessing congested roads because they do not have much money (i.e., they prefer to pay with their time—being stuck in traffic). Given the large degree of income inequality in US and Chinese cities, tricky political issues arise concerning whether a large majority of citizens would support such road pricing. To economists, the key issue here is what becomes of the collected revenue. If the revenue generated by road pricing can be credibly directed to providing public goods that the poor value (such as more public investment in young children), it is more likely that a consensus could be formed behind supporting such efficient public policies.

The introduction of a congestion fee for private vehicles at peak commute times in the congested parts of a city would increase public transit demand. This lesson has been learned from the case of the London central congestion charge.[20] Given the economies of scale of public transit use, encouraging the urban middle class to drive less and to use public transit more would offer overall local and global environmental gains.

New Vehicle-Emissions Regulation

An alternative way to regulate vehicle emissions is through new vehicle-emissions control technology. The United States experience highlights the constructive role that emissions control regulations can play in sharply reducing the urban pollution from vehicles. The US Environmental Protection Agency started to regulate new vehicle emissions in 1972, and over time the regulation has tightened so that vehicles built in recent years are 99 percent cleaner than were new vehicles built in the preregulation period.

This regulation has not come without a cost. The economics literature has debated how such regulation has affected the profitability of carmakers and the purchasing power of car buyers as the price of vehicles has increased. The conventional wisdom is that this regulation has been a net winner for US urbanites. Given that millions of people live in cities, a small improvement in urban smog benefits all in terms of lower disease risk and bluer skies.

China's vehicle emission standards follows European Union (EU), instead of US, standards. China's standards are a few years behind those of the EU, which is behind US standards. This gap between Chinese cars and US/EU cars may be attributed to the cost of abatement technology.[21] But China is gradually catching up to EU standards over time;[22] since 2010, its current national standards follow those known as EU IV, which were adopted in 2005 (EU V standards were adopted in 2009, and EU VI standards will be implemented in 2016.) Beijing and Shanghai, however, have tighter standards: Beijing adopted EU V in 2013 and Shanghai in 2014.[23] China plans to adopt EU V standards nationwide in 2018. We know of no research estimating the cost of vehicle environmental regulation compliance for car producers in the developing world.

Given that vehicles are long lived durables, new-vehicle regulation will only slowly lower overall vehicle emissions. Similar to adding some hot water to a bathtub filled with cold water, if the fleet on the road consists of mostly used vehicles, new-vehicle regulation will only have a significant impact on reducing air pollution as the share of preregulation cohort vehicles on the roads declines.[24]

Gasoline Refining

China's gasoline is known to feature high sulfur content because the refineries have not had sufficient incentives to reduce emissions per gallon of gasoline.[25] Vehicle manufacturers complain that the gasoline is not clean enough to comply with new clean air standards. It is essential to improve the fuel quality (reducing the sulfur content) at the same pace with vehicle quality in order to achieve vehicle emissions reduction goals. The current limit of sulfur content in gasoline is 150 parts per million (ppm) in most Chinese cities, while that in Hong Kong is only 10–50 ppm.

Over the years China's national oil companies have delayed reducing the emissions from the use of diesel fuel. Some environmental scholars have argued that the political clout of SOEs such as oil refineries protects them from attempts to impose environmental regulation on them. The SOE leaders often sit on key government committees that set fuel standards and are thus effectively engaging in self-regulation. In such a case, it is not surprising if weak regulations emerge from the political process. We will return to this point in chapter 8. This discussion suggests that China's environmental progress could accelerate if the national government continues to privatize SOEs. Only in February 2013, after China reached hazardous levels of air pollution, did the State Council issue new guidelines that called for a nationwide adoption of the new China IV diesel standards by the end of 2014,[26] but is yet to be seen whether national oil companies will comply with these standards. The *Wall Street Journal* has reported that such oil refineries will have some ability to raise gas prices to share these new regulatory costs with gasoline consumers.[27]

Smog Tests for Older Vehicles

Given that new vehicles emit less pollution than do older vehicles, old vehicles account for a disproportionately large share of vehicle emissions. It is claimed that in the United States 90 percent of local emissions from vehicles is produced by just 10 percent of the vehicles. Such "superemitters" can be identified using stringent smog checks. In Chinese cities, all vehicles are required to have an inspection every two years, but older vehicles (those older than ten years) are required to have an annual inspection. Such inspections require investments in emissions control technology if the vehicle fails the smog test.

The *New York Times* has reported that in the name of reducing urban air pollution the Chinese government is requiring millions of older vehicles to be taken off its roads.[28] In the case of the United States, there are few vehicles over the age of fifteen driven on the roads. As vehicles age, they are often shipped to poorer nations. International trade in used vehicles between rich and poor countries functions as an informal "cash for clunkers" program: wealthy countries send used cars to poor countries, and poor countries send cash to wealthy countries.

Vehicle Quantity Restrictions

Around the world, a common private vehicle policy that has been in adopted in nations ranging from Brazil, Chile, Colombia, and Mexico to China is to ban driving of certain vehicles on certain days based on one's license plate number's last digit.

Beijing first implemented this policy during the 2008 Olympics. At that time, if the last digit of your car's license plate was an odd number, you could drive on the odd calendar days, or with an even number you could drive on the even calendar days. Economists have been quick to point out the unintended consequence that people buy a second car, and emissions will go up if that second car is older.[29] This policy unintentionally creates an incentive for households to acquire additional vehicles. A driver with two vehicles can drive every day of the week as long as the last digits of the license plates are different.

The Beijing municipal government initiated a new car-control policy when it saw that the driving ban policy was not effective. Since 2011, all car buyers have to obtain a quota through a lottery system. This lottery system limits the number of new car licenses in Beijing to 240,000 per year. This rigid restriction directly controls the number of vehicles newly registered in Beijing. Only those with driver's licenses but without a car can participate in the lottery, which runs every month. Siqi wants a new car so that her family can overcome the driving ban policy. She already owns a car, so she cannot enter the lottery, but her husband can; he registered in the lottery system in 2012, but as of the time of this writing his number had still not come up. A graduate student at Tsinghua who has a driver's license but did not have a strong incentive to buy a car also registered in the lottery just for fun. She was very lucky and received one quota in the month after registering. If she had given up this quota, she would be forbidden to reenter the system for two years. Understanding these rules, she decided to buy a small car and now drives it to campus. In contrast to Beijing, Shanghai has been auctioning such licenses. Li Shanjun reports that in 2012, the Shanghai auction system generated over 6.7 billion yuan to Shanghai's municipal government. The average bid for a license reached over 92,000 yuan in March 2013, higher than the price of many entry-level vehicle models.[30]

Subsidizing Green Car Production

In recent years China's government has made major investments in research breakthroughs to produce low-polluting, high-fuel-efficiency vehicles. Some Chinese cities have made enormous investments in producing and developing electric vehicles. For instance, Wuhan, the capital of Hubei Province, announced that it will subsidize firms producing electric cars or new-energy cars with inexpensive land, low interest rates, and tax cuts.[31] This favorable policy has attracted several electric carmakers such as Dongfeng Renault and Putian. Economists often question whether government officials are able to "pick winners" and wonder whether such an activist industrial policy is a good use of public funds. One distinction here is that the electric vehicle industry is an example of an industry that if it does succeed and displace conventional vehicles this would offer social benefits in mitigating local and global pollution externalities. This is one intellectual justification for such targeted green subsidies. It remains an open empirical question whether the subsidies that California is offering electric vehicle companies such as Tesla, and the similar investments made by China's cities, will turn out to offer private economic returns to these geographic jurisdictions and raise the probability of the development of a viable electric vehicle that middle-class people around the world could purchase.

This matters because the environmental impact of the development of electric vehicles hinges on the likelihood that carmakers develop a new vehicle that could earn a large worldwide market share, and on the source of the power used to generate the electricity that the vehicle runs on. If coal-fired power plants are used to generate the electricity, the deployment of such vehicles could raise aggregate greenhouse gas emissions, while if renewable power is used to generate the electricity, the rise of electric vehicles would be highly beneficial.

Green cars have not yet succeeded in becoming a viable alternative for most Chinese households. During our interviews with Chinese urbanites, some expressed doubt about the ability of green cars to become popular soon. We heard several answers along the lines of, "If I have to purchase a car in the future, I will choose a cost-effective car model and will not pay attention to green cars because I think that they have a low safety performance." Another interviewee offered a different reason for why he would not consider purchasing such a car: "They are not convenient, since you have to charge them. Then, if they are more expensive and less convenient, why should we buy them?" In the United States, many electric vehicle owners have private garages and recharge their cars there. In China, people live in large apartment or condominium complexes, and few of the local parking spots in them are equipped with electric vehicle rechargers. If the price of gasoline were to rise, or if electric vehicle technology improves, more real estate complex developers will invest in recharging equipment for their parking facilities.

It is relevant to note that in early 2014 the electric vehicle maker Tesla announced that it is increasing its production of Teslas in China to sell there.[32] The future of the deployment of the electric vehicle in China hinges on the relative prices of electricity and gasoline and on technological progress in producing high-quality vehicles. If electric vehicles gain an increased market share, and if the electricity is generated by renewables, China (and the United States) will achieve the win-win of freedom of private transportation without the negative environmental effects of local air pollution and greenhouse gas production.

Now that we have examined the economic geography of industry, population migration both across and within cities, and the rise of car use in China, we have completed our investigation of the supply of urban pollution. In chapter 6 we being our discussion of the demand for blue skies, and argue that increasingly well-educated Chinese urbanites have a strong taste for less pollution and less risk in their daily lives.

PART II

The Rising Demand for Green Cities

CHAPTER SIX

The Rising Demand for Blue Skies and Urban Risk Reduction

Pollution poses severe risks to Chinese urbanites' health and quality of life, and as they grow wealthier, they are increasingly willing to pay to reduce their exposure to pollution. All people seek to be safe, healthy, and comfortable, and pollution threatens one's ability to achieve these goals. Wealthier and more educated urbanites have the resources and the know-how to recognize this. Around the world, we observe wealthier people living in safer neighborhoods, eating higher quality food, and accessing better health care than do poorer people.

Risk reduction makes daily life less stressful and increases overall life expectancy and life satisfaction. Death rates in China's cities are declining due to improvements in medical care, better diets, and declining smoking rates. The infant mortality rate in Chinese cities has decreased sharply, from 1.73 percent in 1991 to 0.58 percent in 2010. Life expectancy at birth (in years) improved from 66 years to 73.3 during this twenty-year period. Despite this progress, exposure to China's urban high levels of particulate air pollution are associated with decreases in average life expectancy and higher rates of cardiorespiratory mortality. This finding suggests that air pollution may be an important factor in explaining China's modest growth in life expectancy relative to other nations such as South Korea and Taiwan.[1]

In discussing the demand for blue skies we divide the population into three age groups—children, adults, and the elderly—and investigate their respective quality-of-life concerns in urban China today.

Children

Reducing children's exposure to air pollution goes hand in hand with learning in school and achieving one's full potential. To appreciate this point, consider Jessica Wolpaw Reyes's research, which seeks to understand why crime in US cities increased from 1970 to the early 1990s and then declined sharply.[2] Reyes's empirical research points to the rise and decline of leaded gasoline use in the United States! In the early 1950s, driving in the United States was sharply increasing as more Americans were buying cars and using leaded gasoline. In aggregate, such lead emissions raised ambient lead concentrations in cities. Public health research has documented that lead exposure lowers a child's IQ and increases the likelihood that the child will have attention deficit disorder. Such affected children are more likely to engage in criminal activity as young adults (i.e., twenty years later). This dramatic example highlights the connection between early life exposure to pollution and later life outcomes. With the phasing out of leaded gasoline in the mid-1970s in the United States, the new cohort of young children were exposed to much less pollution and by the time they were young adults in the early 1990s, crime started to fall. Of course, this ingenious theory is not the only reason why crime went up and fell in the United States, but Reyes documents that this theory has greater explanatory power than other explanations such as legalized abortion.[3]

There are two ways to interpret Reyes's findings. A pessimist would say that early exposure to pollution sets off a type of domino effect such that a person never recovers from the initial scarring caused by the exposure. An optimist could reinterpret this lead example to mean that reductions in early life exposure to pollution could have large long-run benefits because healthy kids learn more in school and are more likely to achieve their full potential.

This dynamic vision of compounding investments over the life cycle dovetails with the Nobel laureate James Heckman's highly influential work on skill formation during one's lifetime; he emphasizes that "learning begets learning and skill begets skill." Parents who invest their time and resources early in a child's life raise the likelihood that this child will be more productive in learning at every stage of development. Unfortunately, some parents do not invest time and resources in their children's early education, and such differential investments across parents translates into large differences in children's vocabulary, ability to think creatively, and develop noncognitive skills. Heckman argues that the rise in US income inequality across adults can be traced to differential investments in these adults when they were very young. This research finding has crucial social implications. A newborn child does not choose her parents, and a child who suffers from underinvestment on the part of her parents is less likely to achieve her full potential; the child and society as a whole suffer from this underinvestment. Both for equity and efficiency reasons, Heckman makes the compelling case that society must supplement early investments in children through programs such as prekindergarten education to help every child achieve full potential.[4] Such an investment would simultaneously reduce income inequality in the United States and would raise the nation's future productivity (and hence enhance future economic growth).

This research agenda is highly relevant for China today: if Heckman's core hypothesis is correct, Chinese children who are exposed to high levels of pollution may never achieve their full potential. Exposure to pollution increases sickness and reduces a child's ability to learn and grow. In this case, China could be unintentionally slowing down its overall human capital because a nation's human capital is ultimately determined by the investments made in individual families. From a macroeconomic growth perspective this is crucial, because modern economies such as South Korea and the United States mainly grow due to the accumulation of human capital. Growth theorists emphasize the "law of seventy-two": A nation whose per capita income grows by 3 percent per year enjoys a doubling of its income in 24 years while a nation whose per capita income grows by 8 percent per year enjoys a doubling of its income in nine years. If human capital is the engine of modern urban growth, and if exposure to pollution inhibits human capital accumulation then reducing exposure to pollution becomes a macroeconomic priority for achieving long-run economic growth.

Urban Chinese parents are aware of the channels that Heckman has highlighted. China's unique one-child policy creates the incentive for households to invest more time and resources in children.[5] Known as the "little emperor" or "little princess," the one child is the focus of the modern urban family. After coauthor Siqi Zheng's son was born, her family's monthly expenditures doubled. She imports milk powder from Japan, buys toys, employs a daytime nanny, and is renting an extra single-room apartment in the same community for her parents, who now take care of her son. She has also bought an expensive indoor air filtration machine to protect her son from the Beijing haze.

Since China's tainted baby milk scandal in 2008, more than 60 percent of the baby milk Chinese parents buy comes from overseas, and this brings a 33 percent price premium.[6] Such a premium indicates that urbanites are sacrificing other consumption in order to produce healthy children. One of our interviewees, Ms. Wang, explains that domestic products are not trustworthy, so she asks her relatives who live in Germany to buy milk powder for her. Like Ms. Wang, all of the parents we met during our interviews said that they buy milk powder from abroad.

Our recent research has examined how household Internet purchases are affected by breaking news about food safety. By collecting the sales data of both domestic and foreign brands of milk powder on China's largest online-shopping website, Taobao.com, we find that when a piece of negative news regarding a brand's milk powder quality is reported on the Internet, its sales drop significantly. We also find that over the last three years there were many more negative pieces of news about international brands than Chinese domestic brands. This may reflect the central government's intended strategy: it wants to protect domestic milk makers because such producers generate more jobs and economic activity for the Chinese economy. As Chinese parents of young children seek out safer milk, how will the market respond? Free-market optimists might posit that companies now have a greater incentive to earn a good reputation for producing high-quality, low-risk milk because there are a growing number of sophisticated Chinese consumers who are searching for objectively less risky milk and will "vote with their pocketbook" and not buy milk that they perceive to be risky. If such companies anticipate that they can use bankruptcy to protect themselves when a milk scandal arises, they will have weaker incentives to take costly actions (such as monitoring input providers) that reduce milk risk.

Juanfeng Zhang, an associate professor at Zhejiang University of Technology, pays special attention to the environment his seven-month-old daughter is exposed to. At home he cultivates flowers and plants to improve the indoor environment quality, and takes steps to prevent dust from entering through the windows.

Yang Zan is an associate professor of real estate at Tsinghua University. Before she joined Tsinghua in 2009, she lived in Sweden for several years. Since coming to Beijing, she takes her son back to Sweden for a whole month every summer so that they may breathe fresh air.

Adults

Educational attainment in major Chinese cities is rising sharply over time. In the year 2000, 11.0 percent of urban Chinese people over the age of twenty-five years old had at least a college degree. By the year 2010, this share had increased to 21.2 percent.[7] Better-educated people are more likely to value living and working in a clean environment; based on data from the United States, the better educated are often more willing to pay higher taxes to support regulations intended to enhance environmental quality. In several of coauthor Matthew E. Kahn's past academic studies, he has used voting data from California to explore who votes "for the environment." California offers its voters the opportunity to enact new laws based on direct democracy. During each election cycle they are asked to vote on specific binding ballot propositions. In recent years, Californians have voted on many environment related issues, such as continuing the construction of California's high-speed rail system and introducing new measures to sharply reduce the state's greenhouse gas emissions. While no one individual's vote is observed, the count of yes votes by voting precinct are collected. Matt has used these data, combined with geocoded census data, to study which types of communities are more likely to vote yes on environmental initiatives. A consistent finding in this research is that more educated communities, and communities whose employment is unlikely to be affected by the legislation, are more likely to vote yes on introducing more environmental regulation.[8]

One explanation for why the educated are more likely to be environmentally conscious is that education causes individuals to be more patient and better able to imagine the future. Such individuals may be more willing to sacrifice current consumption, and thus to pay pollution taxes, in return for future quality of life. Better-educated people are also likely to be able to better understand complex phenomena such as climate change and to appreciate the "big picture." Since many environmental science issues are highly complex, the better educated are better able to appreciate the nuances posed by emerging threats such as climate change and biodiversity loss.

While China does not hold direct elections, we are confident that the relationship documented in the United States also holds. Chinese urbanites have other options to express their demand for higher quality of life, such as commenting via social media. We have interviewed several well-educated people during the writing of this book, and a large number of the interviewees have a green-oriented focus. Most of them are currently working in Beijing, for professional purposes rather than due to its quality of life; they are aware that environmental quality is relatively poor there, but they reveal a progreen attitude. "I support all kinds of ecofriendly behaviors, because the environment is a long-term issue. We should give careful considerations to these issues instead of pursuing short-term economic development," says Ms. Lin, a sales manager in a real estate investment company. From our interviews, we hear that people are taking concrete actions such as saving water and electricity and taking the subway instead of driving a car. Such individuals are also actively involved in terms of their Internet reading activity as they forward and comment on articles related to environment degradation and discuss these issues with their friends and relatives. Ms. Zhong lives in the first energy-saving residential building project that focuses on green technologies and low energy consumption in China. "When my twenty-three-year-old daughter came back from a study period in Europe, she put four bins for waste sorting in our house. At that time, such waste sorting was not conducted in trash removal in Beijing. We are now paying attention to environmental protection," she comments.

In Beijing, wealthier residents are not satisfied with local air quality. Though they have a greater preference for clean air, they are not able to control the outdoor air quality in the city, which depends on many external factors such as polluting factories, traffic, winter heating, and climate. Among our interviewees that were members of this new emerging upper middle class in China, all of them declared their willingness to spend money to protect themselves and their families from air pollution. Wearing masks when air pollution is serious has become common. Some are moving to cleaner places. For example, Ms. Zhong decided to live where she does because of the building's focus on green technologies and low energy consumption: "The high technologies in our buildings keep my house at a constant temperature and reduces humidity so that I do not have to crank up any air-conditioning or electric heater. I feel comfortable all year long." Though she paid a higher price, she is confident about the return on this investment since her house saves a great deal of energy thanks to an efficient thermal insulation system. Mr. Wu, whom we first mentioned in the introduction to this book, is even considering moving to Canada or the United States so that his daughter can enjoy better environmental quality.

Many adults in Chinese cities are changing their habits as they attempt to live a healthier life. Several senior colleagues in Siqi's department had been smoking for three decades. They said smoking could help them to be more focused when they were doing research or preparing teaching materials. But in last year they all abandoned smoking after one colleague was diagnosed with lung cancer. (He was cured after a half year's treatment.) All faculty members cheered for this great change. China has over three hundred million smokers, and more than a million Chinese die from smoking-related illnesses every year. Beijing began imposing the country's toughest ban on smoking in public places on June 1, 2015; it bans lighting up in restaurants, offices, and on public transportation. After this ban, many of Siqi's friends in Beijing want to quit smoking because there is nowhere to smoke.

The Elderly

Siqi's father is seventy-six years old and her mother is seventy-two; both are retired high school teachers. They are healthy, and they live near Siqi, taking care of her young son. As they learn more about health, safety, and the environment, their demand for quality of life also increases. Before they came to Beijing, they had never had their teeth cleaned. Now they have regular teeth cleanings.

Young adults can find it difficult to strike a balance between career ambitions and quality of life. But as people age they place greater weight on their quality of life over their professional career goals. Ms. Li, a professor at a university in Jiangxi, is a wise woman in her fifties who is now focusing more on her comfort and quality of life. She has two nonprimary houses away from Jiangxi Province, in Hainan and Kunming. She said that after she retires, she would like to spend the winter in warm Hainan and the summer in cool Kunming to enjoy their high-quality environments. Many Chinese people are following the same strategy as Ms. Li. Matt is from New York City, and he knows many senior citizens who spend their winters in warm Florida. Such US "snowbirds" would understand Ms. Li's lifestyle choices.

Internet Chat about the Environment

Ms. Ju works for the government. She believes that the Chinese public can signal the government about their environmental concerns using such modern social media as the microblog Weibo (the Chinese equivalent of Twitter) and Weixin (WeChat). While any one individual's tweet is unlikely to cause a government office to focus on a specific challenge, if thousands of individuals voice similar concerns and complaints, the government will notice.

Under president Xi Jinping's leadership, Beijing has increased surveillance and regulation of the World Wide Web. The surveillance mainly targets politically sensitive issues such as the Occupy Central events of fall 2014 in Hong Kong. In contrast to such political issues, the central government imposes few limits and little censorship concerning chat about environmental issues. This fact may be due to the central government's eagerness to monitor local governments' real efforts toward combating pollution. China's new Environmental Protection Law, which took effect on January 1, 2015, requires the disclosure of polluting firms' emission information and encourages public participation in this process. However, some local officials still have the incentive to block the announcement of the dirtiest plants in their jurisdiction because people's protest against those biggest polluters will hurt their political career. Yet with the popularity of the Internet and smartphones, it is becoming increasingly difficult for local governments to hide such facts. One piece of evidence supporting this claim is that the daily announcements of ambient particulate matter (measured as $PM_{2.5}$) in Beijing, as announced by the Ministry of Environmental Protection, are highly positively correlated with the tweet readings provided at the US embassy in Beijing.[9]

This use of social media to pursue risk mitigation objectives in China has already succeeded in moving polluting plants away from cities and forcing the government to publish $PM_{2.5}$ data. "I express my opinion by forwarding and commenting some entries in Weibo and Weixin," Ms. Ju notes. Those who express themselves online tend to be the emerging middle class, however. Pollution in remote areas tends to be ignored because those exposed to it are less likely to use the Web to collectively seek environmental solutions. This might lead Chinese environmental policy makers to place the interests of the vocal urbanites above others. With modern media and such information technologies blogs, microblogs like Weibo, WeChat, and mobile phone messages, the government does not have an information monopoly any longer, though it can observe emerging trends in public opinion and concern. While Chinese citizens do not directly vote, such expressions of concern through social media allow them to inform their government of their priorities. Whether rising concern for the urban environment translates into increased regulation hinges on the incentives of government officials. We will explore these issues further in chapters 8 and 9.

There still exists "hidden information" that the government has not revealed. For instance, some polluting enterprises directly discharged untreated wastewater into the groundwater supply, and this fact was not discovered for a long time. China Central Television reported that 55 percent of Chinese cities' underground water is rated at level 4 or 5, the latter being the highest pollution level). When the public began to become aware of it, the problem attracted huge attention from the media and on the Internet. The government has increased its transparency concerning environmental matters. Similar to the CSPAN cable channel in the

United States, the Chinese Ministry of Environmental Protection now televises hearings related to environmental matters. In the past, these deliberations took place behind closed doors.

Migrant Workers' Urban Quality of Life

We have interviewed a few poor migrant workers, including a dishwasher in a university canteen, a cashier in a supermarket, and a cleaning lady in a dormitory. They unanimously worry about the rising cost of living in Beijing. Even though they benefit from some financial help—for example, the university provides dormitories for the dishwasher and cleaning lady—they are aware that without this help they would have been forced to move out of Beijing because they cannot afford any housing unit. "I am already old, but I work very hard to save money in order to give my son the chance to go to university," notes the dishwasher.

Many migrant workers in big cities must also choose whether to stay or leave these areas, and their choices in aggregate will influence the future of Chinese cities. Recently a new trend known as the *tide of return* has taken place in China.[10] Right after the 2009 financial crisis, millions of migrant workers found themselves without work and were forced to return home. Some of them decided not to come back to the large coastal cities after the economy recovered.

Migrant workers in big cities like Beijing or Shanghai suffer from harsh living conditions. They live in low-quality housing units, sometimes underground, and their children are crammed into low-quality schools. They are not integrated into the society, and often face discrimination and alienation. "We have never quite felt at home in Beijing," a migrant worker explains.

Inland cities are experiencing radical change. During the years those migrant workers worked in first-tier cities, economic development and social welfare reform came to their home regions. This regional convergence is mainly due to the new vision of the Chinese leaders to promote the growth of small- and middle-size cities. Large inland cities such as Chongqing, Taiyuan, and Zhengzhou are becoming attractive places to work for people from the rural parts of those regions. Improvements in the intercity and regional transportation network allow these workers to come back to their hometown on weekends to see their relatives, whereas they used to only visit once a year. These small- and middle-size cities do not suffer from extremely high housing rents, severe air pollution, or congested roads; the quality of life there might be much higher than that of first-tier cities.

Many low-skilled migrant workers choose to return to the small inland cities with more job opportunities, lower pollution, and lower living costs so that they can enjoy an improvement in their quality of life. The most talented stay in the superstar cities and face more stress and high home prices, but also live a more vibrant life. The old cliché is "different strokes for different folks." Facing the intercity real estate price gradient and recognizing the different employment opportunities, individuals will make their own best choices of where to live and work. In a diverse economy, the key is for them to have a menu of options to choose from.

There is a sharp contrast between the lives of wealthier residents and those of the urban poor. The urban poor have lower quality of life because their day-to-day existence often boils down to work; they have very few leisure or recreational activities. They are struggling to satisfy their basic needs, and do not think about quality of life. Siqi's nanny works on most weekend days and only enjoys two days of leisure per month. She prefers such a busy work schedule because she wants to earn as much money as possible, and she has few friends in Beijing to meet with after work. Though she can afford renting a small one-bedroom apartment (around thirty square meters) in Siqi's neighborhood, she rents a very tiny room (about eight square meters) in a poor urban village nearby. "I only need a bed and I do not care about my quality of life in Beijing," she explains. "My first priority is to save as much money as I can, and after that, I will return to my hometown."

Improving Urban Quality of Life Reduces the Risk of a Brain Drain

Many educated people are thinking about leaving China to settle in Australia, Canada, Europe, or the United States, where environmental quality is higher. If China achieves blue skies, it will slow down the brain drain and capital drain caused by elites leaving the country. International corporations will be more willing to locate in Beijing if they do not have to pay "combat pay" to lure talented expatriates to live there. If China's major cities' quality of life improves, high-skilled international people will be willing to live and work in these cities, and this will augment the stock of human capital. Given the importance of face-to-face communication in the modern skills economy, China's economy will grow even faster if it can attract and retain the skilled internationals.

Over the last few decades, tens of thousands of brilliant and ambitious Chinese citizens have studied at US universities, and many of them have then taken jobs in Canada and the United States. If quality of life improves in China, many of them will consider returning home; in fact, 25 percent of Chinese startups registered in Beijing's Zhongguancun Science Park were founded by overseas returnees. This example highlights that improvements in China's urban quality of life will simultaneously reduce outmigration and stimulate inmigration of high-skilled individuals to move to China's cities.

This chapter has provided an overview of the demand for blue skies. In chapter 7 we will discuss some of our recent research attempting to quantify this demand using revealed preference methods.

CHAPTER SEVEN

Recent Empirical Evidence on the Demand for Lower Pollution Levels

Since 2006 we have written several empirical studies examining Chinese urban household demand for improved environmental quality of life. In this chapter we provide an overview of the set of questions that motivated our studies, the data we used, and our main findings. This discussion helps to highlight how economists make research progress on environmental and urban economics issues and pinpoints the still open questions. Our focus is on the demand to avoid local air pollution and the demand to live close to green space.

As economists we seek to learn about people's priorities through the choices they make in markets. If an urbanite pays five hundred dollars more for a condominium with a beautiful view than she would have paid for a similar condominium without such a view, we can quickly conclude that she values such a view by at least five hundred dollars. Why? If she was only be willing to pay four hundred dollars for such a view then she never would have paid the five-hundred-dollar premium. This simple revealed preference logic forms the basis for much of the evidence we will present in this chapter.

In our academic research we have used high-quality real estate transaction data from Chinese cities to explore at a given point in time where home prices are high. High market prices signal quality. After all, if a condominium is of low quality, why would a person bid too much for it? From observing the market price of Chinese condominiums and by using geographic information software to describe the condominium tower's local attributes (such as local air pollution levels or distance to the city center, public parks, or nearest subway stop) we have broken down how much of the price premium for a condominium is due to this bundle of attributes.

Evidence from Beijing

Our first coauthored research paper focused on Beijing condominium owners' willingness to pay for local quality of life based on data from 2006. Every real estate student is taught that the key determinant of real estate's value is "location, location, location," but to an economist this is too vague. On what dimensions do locations differ? One easy metric is the distance to downtown, but there are other location specific attributes such as local air quality and proximity to local parks. We used data on the price per square meter of all new condominiums sold in Beijing during the years 2004–6. For each of these housing condominium complexes, we used geocoding software to calculate their distance to the city center, the closest subway station, major public parks, and public universities. We also collected data by geographic location on the local school quality. At the time we wrote our study, Beijing had a network of over a dozen ambient particulate monitoring stations; we used the annual pollution readings at each of these stations to impute the level of air pollution at each of the housing towers in our data set.

After creating this data set, we used linear regression methods to break down how much of the price premium per square meter of housing was accounted for by each of the factors listed above. Intuitively, this approach seeks to take a standardized condominium (i.e., a unit of a given size built by a specific developer) and estimates what its price would be in different neighborhoods within the same city. By combining information on this price premium across neighborhoods with data on how neighborhoods differ with respect to their local public goods, we measure how much households are willing to pay for nonmarket locational attributes such as the absence of air pollution or the proximity to green space. We find evidence that proximity to fast public transit, clean air, high-quality schools, major universities, and environmental amenities are capitalized into real estate prices in Beijing. These capitalization estimates highlight that Chinese urbanites reveal similar priorities for local quality of life to those of their US counterparts. For example, we found that neighborhoods featuring a level ten units higher particulate matter up to ten micrometers in size (PM_{10}, a key indicator of urban air pollution) are associated with having real estate prices that are 4 percent lower. This implicit compensation for exposure to pollution reveals that real estate buyers value avoiding such pollutants.

Proximity to New Beijing Subway Stations and the Olympic Park

In preparing for the 2008 Summer Olympics, Beijing's leaders made enormous investments in beautifying the city and in improving the city's infrastructure. In a 2013 paper we studied the implications of the building of the Olympic Park and new subway lines on local real estate construction and real estate pricing.[1] The Olympic Park represented a significant investment in new green space, and the new subway lines created new low-pollution, low-carbon mobility options within the congested city. We used this "natural experiment" to study how real estate markets responded to this improvement in local public goods. We knew the exact location of the new Olympic Park and the new subway lines and stations. Using these geocoded data, we documented that real estate prices increased close to the new green-space areas and close to the new subway stations. We also documented that real estate developers responded to these new spatially localized local public goods by building more housing units in close proximity to these new amenities. Finally, we documented the rise of localized "consumer cities" as restaurants and stores such as Starbucks opened up near these new amenities. This evidence all supports the hypothesis that Chinese households seek out a new urbanist lifestyle and that developers recognize this and are responding to these market incentives. While these same points have been documented in the United States (for example, in areas close to new-walk-and ride stations on the Boston subway's Red Line), similar evidence has not been previously documented in a fast-growing developing country.[2]

Intercity Real Estate Price Variation

In recent research we have investigated intercity real estate price differences in China. With the decline of the internal passport system (*hukou*), Chinese households have more choice over where to move. A robust result we find is that real estate prices are higher in those Chinese cities offering such green amenities as clean air, a coastal location, and a temperate climate.

In our research we find that real estate prices are lower in highly polluted Chinese cities, and this discount is growing over time as the average urbanite becomes wealthier and better educated. A novel feature of our study is that we observe intercity variation in pollution because there are sandstorms and factory emissions crossing city boundaries. Many studies seeking to measure the association between local real estate prices and pollution levels face the challenge that pollution levels could be high in a city because the local economy is booming. In such a case, a naive researcher might observe high pollution and high real estate prices and falsely conclude that people *like* pollution! The mistake in this case would be that the researcher attributed all of the intercity variation in real estate prices to local pollution levels, when in fact there was an omitted third factor (whether the local economy is booming) that is the true cause of both the pollution and the high demand to live in that city. One solution to this challenge is for the researcher to identify other factors that determine a city's air pollution levels but are unlikely to be related to the local economy's vibrancy. Sandstorms and cross-boundary industrial emissions meet this threshold; both raise local air pollution levels but are unlikely to be correlated with local economic business cycles. The director of Beijing's Environmental Protection Agency found that 28–36 percent of local pollution came from regional "imports." In a recent study, we exploit the fact that sandstorms affect Chinese cities differently and estimate even larger effects of city pollution on city home prices.[3] Such sandstorms induce variation in a city's ambient pollution that is likely to be independent of other factors in local home prices. In this sense sandstorms allow us to generate more credible estimates of the role of urban pollution in determining real estate prices.

In another intercity study, we investigated the impact of China's bullet trains on real estate price dynamics in second-tier cities located relatively close to superstar cities.[4] As we discussed in chapter 3, the introduction of the bullet train connects cities located 50–300 miles away from major cities so that these medium-distance cities are now roughly one hour away in commute time from superstar cities such as Beijing or Shanghai. We document the jump in real estate prices in such second-tier cities after they were connected to the major cities by bullet train. This finding is relevant for those interested in urban sustainability because it highlights how major infrastructure investments by the Chinese central government helps to create a system of cities and creates a greater menu of residential options from which households may choose. This menu of choices provides individuals with greater opportunities and implicitly protects them if the major city becomes too expensive or suffers from significant crowding and pollution.

Air Mask Sales and Government Announcements

Real estate markets are just one example through which Chinese urbanites express their desire for blue skies. Purchases of devices such as air masks and filters represent a self-protection investment that Chinese urbanites can buy to protect themselves. If a person purchases a mask for the equivalent of ninety cents and this provides one week worth of protection against air pollution, an economist infers that this person is willing to pay at least ninety cents per week to avoid the health risks caused by air pollution. One recent study concludes, based on masks sales in China, that if 10 percent of heavy pollution days in China were eliminated, a lower bound on the value of such an air quality gain would be $187 million dollars. This estimate is based on the extra expenditure on air masks caused by high pollution levels; the masks would not have been purchased if the air had been cleaner, and thus the expenditure on such self-protection provides evidence on the value of air-quality improvements.[5] We provide this example to show how economists use market data to learn about how households value nonmarket goods (clean air). The actual cost of China's urban air pollution is likely to be much larger than what is revealed by expenditure on masks. A joint World Bank–government of China report released in 2007 reports that the combined health and nonhealth cost of outdoor air and water pollution for China's economy comes to around $US100 billion per year—about 5.8 percent of the country's gross domestic product.[6]

Large Chinese cities now report their daily readings of PM_{10} and $PM_{2.5}$ on television and post this to the Internet. In chapter 1 we mentioned the iPhone application displaying air pollution indexes published by China's Ministry of Environmental Protection and the US embassy. Most members of the new emerging class of young educated urbanites in China, like coauthor Siqi Zheng, have this application on their smartphones. Such applications are regularly checked. In another study, we document that Chinese urbanites respond to this real-time information by increasing their purchases of air masks and filters on days when the air is more polluted.[7]

Beijing urbanites check their smartphone applications displaying $PM_{2.5}$ readings based on the standards of the US Environmental Protection Agency.[8] After dozens of Chinese cities began to change the standards and publish real-time $PM_{2.5}$ concentrations in 2013, the local public gradually trusted the official data since their readings were not very different from the data reported by the US embassy.

In our recent research studying Internet sales for masks and air filters, we find that daily sales are responsive to government announcements. Sales of air masks and filters are much higher on days when the government announces that a city's air pollution is "hazardous" versus when government announces that a day's air quality is "excellent." Our findings suggest that, at least recently, China's urban consumers trust their governments' pollution announcements. The local government can help protect its people by providing accurate, up-to-date information about outdoor air pollution; such information allows individuals to reschedule their daily activities to reduce time outside and to invest in self-protection.

Our findings based on air mask sales in China build on a recent literature from the United States that examines how people respond to trusted government announcements about air pollution alerts. Matthew Neidell studies how smog alerts affect household behavior in the Los Angeles area. While ambient ozone pollution readings can take on any value of zero or greater, the California regulators have declared a sharp cutoff such that when pollution exceeds this critical value that a smog alert is announced for the next day. Neidell uses daily attendance at outdoor events in Los Angeles Zoo and Dodger Stadium as indicators; both venues are located in a smoggy part of the city.[9] He documents statistically significant evidence that attendance drops at those events when a smog alert is announced.

Consider attendance at Los Angeles Dodgers baseball games. Throughout the summer the daily temperature and smog level varies; if a great visiting pitcher will be pitching, a large crowd may attend even on a hot, smoggy day. Neidell studies whether attendance at the smoggy baseball stadium is lower on days when a smog alert has been announced. As a control he compares average attendance on those days to average attendance on other smoggy days that are just clean enough to not trigger such an alert. This natural experiment highlights that big-city residents in the United States are responding to public health announcements by avoiding polluted places on dates when the government has announced that pollution levels are high. Our study of air mask demand from China reveals similar patterns. The public's willingness and ability to engage in self-protective steps reduces its exposure to outdoor air pollution and allows individuals to suffer less health risk from outdoor pollution.

Rising Quality-of-Life Inequality in Urban China

By choosing a city, and a neighborhood within that city, urban residents have some control over their exposure to air pollution. Real estate prices are higher in less-polluted geographic areas, and the wealthy tend to live in these areas.[10] Using cross-sectional data on real estate prices across Beijing, We find that a home's price is lower when it is located in an area with a higher average PM_{10} concentration.[11]

The poor also cannot afford cars. Many of them walk or ride bicycles, so they are exposed to more particulate matter during hazy days. Wealthier people buy many more air filters and slightly more masks than do poor people. This means that private investments against pollution risk enlarge the public heath inequality between the rich and the poor.

In the previous section we argued that Chinese urbanites have access to an increasing set of options to help them reduce their exposure to pollution. Some of these products are not cheap, and this means that the poor will be less likely to buy them. An air filter is much more expensive than a mask; their average prices are US$490 ninety cents, respectively. Consumers have to change the air filter's strainer once per year, but a mask only lasts for about ten days. Thus, the daily user cost (including electricity expenditure) of an air filter is more than ten times that of a mask. Using data on sales of different products on Taobao.com in 2013 by city, day, and income category, we find that wealthier consumers are more likely to buy air filters.

This suggests that pollution exposure inequality is likely to be rising in urban China. When the outdoor air is polluted, people (especially the wealthy) reduce their time spent outdoors. According to a nationwide survey in China, high-income people spend 20 percent less time outdoors than do low-income people. Wealthier people are more likely to commute in cars, and this protects them from the dirty air outside. Wealthier people will have access to the best masks, filters, and health care to help them to adapt to pollution.

Such differential exposure to pollution across income groups means that measurement of income inequality at a point in time is likely to understate true inequality because wealthier groups will be likely to have a longer life expectancy and to enjoy a greater quality of life each year.

Further compounding the environmental justice challenge is the fact that government officials are likely to be more responsive to the concerns of the middle class and wealthy in their cities. In the urban United States, poor renter communities often have little voice in the political process, and this has raised the likelihood that such areas are chosen to site new toxic facilities. It remains an open question whether China's urban growing environmental movement will exacerbate or mitigate an inequality between the rich and the poor as regards pollution exposure. The voices online tend to be the wealthy and the emerging middle class; the poor are underrepresented in the public media. In addition, pollution in remote areas tends to be ignored because the people there are less likely to use the Web to collectively seek solutions to environmental problems. Such biases may exacerbate quality of life inequality because environmental policies place the interests of vocal urbanites above others. Standard real estate economics also argues that home prices reflect the quality of the neighborhood; thus, the worst areas will have the cheapest home prices and this will tend to act as a poverty magnet. By attracting the poor to these areas, this further depresses home prices and the process feeds on itself.

Now that we have explored the demand for improved environmental quality, we will turn to an examination of how China's leaders respond to such desires. China is a not a democracy, and in chapter 8 we describe its unique political institutions and explore how citizen preferences influence the choices made by the central government.

PART III

Promoting Environmental Accountability in a One-Party State

The Central Government's Increased Desire to Promote Environmental Sustainability

China's past leaders were not willing to sacrifice economic growth for environmental protection. In the late 1950s, Mao Zedong's five-year plans focused on state-owned enterprises maximizing industrial production. Urban quality of life was never mentioned, nor were concerns about industrial pollution. China's leaders sought to mimic the Soviet Union's vision for how to compete with the West. The Soviet model focused on building a strong military through rapid industrialization. At a time when the Chinese public was recovering from World War II and not well educated, the state faced no pushback from civil society. The paucity of information about day-to-day life limited the ability of dissatisfied individuals to organize and lobby for improvements in their standard of living. In both the Soviet Union and Communist China, the state had a monopoly on information, and the public thus had a limited ability to learn about the unintended consequences of industrialization.

Under Deng Xiaoping's reforms in the 1980s, China unleashed an industrial machine focused on exports to the West; there was an imperative to reduce China's abject poverty. China has achieved an annual 8 percent economic growth for the last thirty years. Until recently, neither national government officials nor local urban officials prioritized environmental protection; China's central government focused on economic growth with an emphasis on gross domestic product (GDP) as the key evaluation criterion for ranking local officials. Such officials sought to boost their local economy by attracting industries and investing in transportation infrastructure projects, but had little incentive to reduce energy consumption or protect the environment in their own jurisdictions because such actions did not help their political careers. Under this old regime, local leaders were more likely to be promoted if they invested in transportation rather than environmental protection.[1]

The Chinese central government is powerful. Past decisions such as the agricultural consolidation under Mao, the one-child policy, the building of the Three Gorges Dam, and free winter heating for citizens, have had profound social and environmental implications.

In recent years a regime shift has taken place so that environmental and energy efficiency criteria are explicitly incorporated into the nation's performance targets. The central government now emphasizes "harmonious development."[2] The State Environmental Protection Administration (SEPA) was established in 1998 and was renamed the Ministry of Environmental Protection (MEP) in 2008.

The Chinese state has established a number of targets for energy efficiency and pollution reduction in the tenth (2001–5), eleventh (2006–10) and twelfth (2011–15) five-year plans (FYPs). At the Copenhagen Climate Summit in 2009, the nation pledged to achieve a carbon-intensity reduction of 40–45 percent by 2020.[3] Such a target is defined as the ratio of carbon dioxide emissions per yuan of economic output. Thus, this ratio can decline even if total carbon dioxide emissions are rising. In the tenth FYP, the target stated that major water and air pollutants should decrease by 10 percent over the five-year period. In the eleventh FYP, the target goal was that sulfur dioxide, as well as major pollutants detectable in a chemical oxygen demand (COD) test, which is commonly used to indirectly measure the amount of organic compounds in water, should decrease by 10 percent each year from their 2005 levels; energy consumption per unit of GDP was also to decline by about 20 percent from its 2005 level. The Chinese central government set a target in the twelfth FYP of reducing energy consumption for every ten thousand yuan of GDP by 16 percent by 2015 (or 3.2 percent per year). In terms of carbon dioxide, China has vowed to reduce the intensity of its emissions per unit of GDP in 2020 by 40–45 percent compared with 2005 levels.[4] Although environmental targets had been incorporated into national FYPs as of the late 1990s, they were a low priority relative to achieving national economic growth goals.[5] In 2006, at the beginning of the eleventh FYP, Beijing upgraded a number of environmental targets from "expected" (*yuqixing*) to "binding" (*yueshuxing*) status. Binding environmental targets were thereafter written into local leading officials' annual responsibility contracts and became important criteria in their decisions regarding their promotions.[6] The intent was to incentivize officials at each government level to fulfill the central government's environmental mandates.[7]

Understanding the Regime Shift

Why has this green regime shift taken place? One might conjecture a "superman theory" that the new leaders of China have stronger preferences for blue skies than Mao or Deng did. If this claim is true, it should make environmentalists nervous, because it would suggest that any current environmental gains could vanish moving forward if China's future leaders have stronger preferences for further industrial growth over environmental protection.

A second theory posits that the Communist Party seeks to boost its own political legitimacy in the international arena as well as with the Chinese people. Making a commitment to pursuing environmental goals is one way to credibly signal to both domestic constituents and international actors that China is an international leader and that the Communist Party leadership cares about its own people.[8] As the nation enjoys the benefits of economic growth, it aspires to be an international superpower, and a superpower wields both hard and soft power.

As China invests billions in its military (hard power), it also seeks to invest in developing its soft power. While soft power is a squishy concept, it can be thought of as representing a capital stock of goodwill and trust earned through past actions. For example, a nation that takes costly actions (such as providing food relief after a natural disaster) will gain no short-run benefit, but its deeds will be noted and this will build international respect. A "leading nation" plays an active role in international relations, helping to keep the peace and promote global public goods. If

China continues to be perceived as the world's environmental villain, this chips away at this aspiration. The nation's nascent efforts to improve its environmental performance may represent a serious investment in soft power as it seeks to elevate itself to being globally recognized as one of the world's leading superpowers.

A third explanation for China's increased focus on goals for sustainability is that the central government believes that the rest of the world is embracing the low-carbon energy agenda and creates a market imperative for China to become a technological and economic leader in this nascent field.

The central government's green focus also reflects the aspirations of its growing urban middle class. For more than three decades, China's economy—and its rapid growth—was the central issue for enhancing the political legitimacy of the country's leaders. So long as GDP grew by double digits, the majority of the people were happy. That's no longer the case. People are increasingly distressed by the accelerated risk of lung cancer caused by heavy smog, by tap water being considered unsafe in most cities, and by young mothers being forced to buy baby formula from abroad. Many educated people have voted with their feet, choosing to move to safer places, and many Chinese have sent their children abroad. As we discussed in chapters 6 and 7, a growing share of China's public demands environmental progress.

While China does not hold direct elections, a government that seeks to remain in power has strong incentives to satisfy the desires of its urban constituents. Throughout the world we have witnessed revolutions in nations whose governments have not delivered what the people expected. While China has experienced amazing growth in recent years, many cities suffer from a low quality of life, and promoting environmental issues may be a low-cost way to build loyalty among the people. This suggests that environmental protection represents a type of double-bottom-line investment that simultaneously improves the nation's quality of life and boosts popular support for the government, hence reducing the risk of social instability.

International research has highlighted that as nations grow wealthier they increase their environmental regulations. This theory predicts that middle-income nations start to increase their enforcement of environmental regulations and these tighten further as more economic development takes place. A classic example of this is the quantity of lead in gasoline, where there is a sharp negative correlation between grams of lead per gallon of gasoline in a nation and that nation's per capita income. As nations grow wealthier they tend to enact and enforce regulation to require lower emissions per gallon of gasoline.[9] An open question concerns whether democracies and nondemocracies are equally likely to increase their efforts at providing public goods such as clean air as economic growth takes place.[10]

How Can the Central Government Encourage Local Mayors to Pursue Green Goals?

Throughout this chapter we avoid the naive view that the central government speaks with one voice. The government consists of many ministries that have different priorities and constituencies. For example, the National Development and Reform Commission (NDRC) has priorities that are different from those of the MEP. The NDRC itself is in charge of multiple functions; its main concern lies in overseeing planning and investment management to boost the nation's economy, while at the same time it is also responsible for overseeing energy efficiency and climate change issues. The MEP has a lower bureaucratic status than the NDRC, and such interagency competition raises issues of who will prevail in enacting and enforcing environmental regulation. We seek to understand how urban growth, rising education, and rising income affects this interest group struggle. If progrowth interests are able to throttle environmental regulatory efforts, it is less likely that significant environmental progress can be achieved.

The central government has unique powers for shifting China's economic geography, and this can have large environmental implications. The 2008 Olympics in Beijing offer one example. In order to make a positive impression on the world television audience, the central government ordered the closing of local factories and power plants to improve air pollution levels. Empirical research has documented that these interventions were effective and that a fairly large share of the pollution gains persisted after the Olympics ended.[11]

China has a strong one-party central government, but hundreds of local governments act as competing enterprises. The central government's rules and institutions play a key role in determining whether it can bring about fundamental changes in quality of life in China's cities. Institutions such as the courts system, the promotion system for evaluating urban mayors, and environmental governance play a central role in contributing to quality of life.

Unlike in the West, where competitive elections are used to allocate government positions, the upper-level governments—rather than voters—decide on the appointment and promotion of lower-level officials. The State Council appoints the governors of provinces, municipalities, and some major cities (so-called provincial-level and vice-provincial cities) directly. Provincial governments appoint the governors of prefecture-level cities. How such officials are selected and promoted is central to the effective governance of the nation. The selection and promotion process is performed by the upper-level Communist Party Committee's personnel department, which is a key sector in the upper-level government. This performance evaluation is based on objective and quantitative targets, individual interviews, and qualitative assessment of candidates' capacity and potential.

Quantitative performance evaluation is becoming more important because it is relatively easy to perform, and it is regarded by the local leaders as a fair way to compare their performance with that of their peers. In the past, GDP growth was the main criteria used by upper-level governments in evaluating the performance of lower-level officials' performance.

Including environmental targets in the performance evaluation and promotion criteria of local government leaders is the state's key approach to address the "principal-agent problem" that arises because the state (the principal) cannot directly monitor the local government leaders' (the agent's) efforts in protecting the environment. Since the eleventh FYP, the state has started to incentivize local governments by including energy conservation and pollution reduction targets in the performance evaluation system of local officials.[12]

The city leaders now are increasingly likely to prioritize environmental issues.[13] China's mayors have always been evaluated for promotion based on their city's per capita income growth and whether there is social unrest. In recent years, the central government has introduced additional performance standards that include reductions in the energy intensity of the city's economy and progress in cleaning up ambient air pollution. The weight placed on these environmental items is fairly large, at roughly 20 percent of the criteria.[14] In our recent research, we have studied the probability that mayors are promoted and found that there is a positive correlation between a city's economic growth and the mayor's propensity to be promoted but, controlling for this factor, the mayor is also more likely to be promoted if the city's energy intensity is declining and if ambient air pollution is declining.[15] This effect is more statistically significant in the large cities of China's more developed coastal region.

The introduction of binding environmental targets has also generated some undesirable outcomes. A potential problem with this evaluation system is that the main targets are sometimes not closely linked to environmental outcomes that have significant impacts on the public's health and quality of life; instead, they are linked to accounting indicators such as energy intensity and environmental infrastructure investment because those accounting indicators can be more easily measured and collected. For example, credit for pollution reduction might be granted for the construction of a waste gas treatment plant or installation of pollution control technology in a power plant. Local officials are thus incentivized to invest in environmental infrastructure and pollution control technology. With insufficient monitoring by the central government, there is much less focus on whether these local investments are operated properly such that they actually reduce pollution. Some factories have adjusted pollution control equipment to report false data, treatment plants have been left idle, and some local governments have forced emergency shutdowns of electricity to local public services to meet energy efficiency targets.[16] An optimistic response is that if the central government recognizes that some local officials are engaging in such shortcuts, it can introduce random audits of such cities and factories to reduce such malfeasance.

Another problem is that the distribution of national targets to local levels is not based on environmental science. This decision-making process requires a constant flow of high-quality information in order to identify the "right" target level for subordinate governments. In the absence of such information, the use of one-size-fits-all targets can distribute the implementation burden quite unequally. For example, within the same municipality in Hunan, one county's Environmental Protection Bureau reported that air pollution targets were "easy" to achieve while two neighboring counties felt they were "difficult." This will supply incentives for heavily burdened localities to misreport data on difficult targets.[17]

The economic approach for addressing this issue would be to introduce a pollution permit market in which there would be an aggregate pollution target that could not be exceeded. Factories that pollute as a by-product of manufacturing would need to purchase a pollution permit in order to operate, and the central government would randomly audit to make sure the firms are complying with the rules. By adopting this new incentive system, polluting firms would be incentivized to reduce their emissions and to invest and operate low-pollution technology (because they would then need to buy fewer permits). Such a market would stimulate research and development on the part of environmental engineers for more such equipment, and this would further reduce the aggregate cost of achieving the pollution-control goal. Note that in this case, the central government would not ask cities whether their factories were having trouble achieving the pollution target; the factories would reveal this information by how many permits they buy in the market. For example, if a permit costs three hundred yuan per ton and we observe a firm buying such a permit, we can immediately infer that this firm would incur a cost greater than three hundred yuan to reduce its emissions by a ton. Why? Because we see it purchase the pollution permit at a price of three hundred yuan when it could have reduced its own emissions. If such a cost-minimizing firm could have hired an engineer for two hundred yuan to reduce its emissions by a ton, it would not have purchased the more expensive permit.

As pollution permit programs are piloted in both the United States and China, government officials in China are likely to grow more confident in the use of such new markets to allocate scarce resources.[18] A skeptic might counter by pointing to the recent failure of the European Union's cap-and-trade market, but the issue that arose in that case was the European nations auctioning off too many permits, which meant that the price for polluting was set very low.[19] A nation that is committed to truly changing the rules of the game and introducing a credible pollution permit market would auction off a large number of permits at first (so that the market price of a permit was low) but then commit to auctioning off fewer permits each year so that polluters would expect rising prices for permits. California is pursuing this strategy as it ramps up regulations in its Global Warming Solutions Act of 2006. Chinese delegations have been flying to California to learn the details about how that state (as a "green guinea pig") is pursuing the sharp decarbonization of its economy. Ideas are public goods; if California succeeds in piloting its program, China can mimic the state's success and avoid the mistakes it makes in the process.

An Example of How Changing Promotion Incentives Improved Water Pollution

Along many rivers around the world, researchers have noted that water pollution is higher at locations close to political boundaries. Such "free riding" takes place because local officials seek the benefits of economic activity (such as having pulp and paper mill production) but also seek to spread the environmental costs to downstream neighbors. Such activity leads to elevated levels of water pollutants such as those found in the COD test. In China, the same dynamic has played out at provincial boundaries. Concerned about high levels of COD pollutants at such boundaries, China's central government changed its political promotion rules to punish local officials for excessively high levels of these pollutants. The government built and maintained a water monitoring system and used data from it to monitor local leaders' environmental performance. Matthew E. Kahn, Pei Li, and Daxuan Zhao have studied this natural experiment in which the central government changed the local political promotion criteria and hence incentivized local officials to reduce border pollution along specific criteria;[20] they document evidence of pollution mitigation progress with respect to targeted criteria at province boundaries. This example shows that when the central government introduces new promotion rules that target a specific environmental outcome, local officials respond; this highlights the importance of choosing the "right target." In the case of water pollution, there are other measures of water pollution, such as heavy metals that may be more important for public health than exposure to those found in a COD test.

Competing Interests within the Central Government

In the Chinese planning system, fuel prices, electricity prices, and related subsidies are set by the NDRC in Beijing, as are the energy-efficiency targets to reduce the energy/output ratio and the emission/energy ratio. The NDRC has resisted imposing regulation that protects the environment, but raises the cost to state-owned enterprises for producing gasoline and electricity.[21]

Consider the case of regulating gasoline in order to lower its sulfur content. The key issue that has arisen is who pays for this costly regulation that will require refineries to invest in changing their production process. The refineries would seek to raise their gas prices to cover their higher costs, but are inhibited by the government's rigid price setting. Urban consumers seek gas price subsidies and cleaner gas, while the gasoline producers would prefer to pass their regulatory costs on to end consumers.

The MEP—the main government advocate for both higher fuel standards and cleaner automobile technology—is relatively weak compared to other central government agencies. It has the power to force automakers to use new technology by issuing stricter tailpipe emissions standards, but it cannot unilaterally impose new fuel standards or enforce compliance from the oil companies. Instead, it can merely lobby other relevant ministries or agencies to take action. When fuel standards do not keep pace with vehicle technology, the MEP has to delay issuing new tailpipe emissions standards.

The MEP was established in 2008, but the environmental protection functions of China's central government have long been dispersed among the NDRC, the Forestry Administration, the Ministry of Water Resources, the Ministry of Housing and Urban-Rural Development, the Ministry of Land and Resources, the State Oceanic Administration, and other ministries and administrations. All of these institutions work in collaboration with the MEP, but their tasks are not clearly demarcated. Their respective functions are represented in table 8.1.

In all of these governmental institutions there can be conflicts between economic profit and environmental benefits. Without the intervention of higher institutions such as national leaders or the NDRC, these lower institutions, as well as local governments and companies, would ignore environmental regulation or even authorize economic activity that creates pollution. The MEP cannot directly intervene in environmental issues managed by other institutions; this hinders coordination and leads to poor management of problems. In 2012 some Beijing residents protested against the noise pollution generated by the Beijing–Shenyang bullet train construction project and blamed the Ministry of Railways for having performed their own biased environmental assessment. The MEP did not have enough power or influence on this issue.

The MEP's influence is likely to grow. This book's coauthor, Siqi Zheng is a professor at Tsinghua University, which is known as a leading center of science and engineering research. In January 2015 it was announced that Tsinghua's president Chen Jining would be appointed as the new head of the MEP.[22] Chen was the former dean of Tsinghua's School of the Environment; his research expertise is environmental policy and management. In her capacity as the director of Tsinghua's Hang Lung Center for Real Estate, Siqi has often met with Chen, and he was an active participant at the Hang Lung Center's last board meeting. In his meeting with Siqi, he expressed interest in our finding of the Chinese people's rising demand for quality of life (including environmental quality). The appointment of a star academic to a major government post reveals that the Chinese leadership sought a top mind to promote environmental progress; if it had simply sought a new professional politician, it would likely have promoted somebody else. Chen has a sterling academic record focused on environmental engineering topics,[23] and his training and background suggest that he is an ideal appointee for helping the MEP grow and succeed in the interest-group struggles that we discuss herein.

Since the establishment of the MEP the media have been concerned about interagency cooperation. The central government declared its intention to integrate all environmental protection functions into the MEP, but solving this decentralized management issue will take a long time. The state knows that the collaboration between a weak institution (the MEP) and a powerful one (the NDRC) is the key to effectively promote the development of environmental protection in China.

These issues of coordination arise in many contexts. Past industrial parcels often feature contamination and thus require remediation investments to convert them into green parcels that can be developed in a housing or commercial tower. Such "brownfield" investments are administered by the Ministry of Land and Resources, which has prevented the MEP from addressing this issue for many years. These clashes across governmental agencies are caused both by the pursuit of power and conflicting objectives. For instance, the Ministry of Railways set as its ambitious goal to construct the longest high-speed railway in the whole world, while the MEP cares more about the environmental consequences of such a major infrastructure investment.

Agencies such as the NDRC have focused on investments that yield economic growth while not paying close attention to environmental impact. A key role that the MEP will play moving forward is to integrate predictions of likely environmental impacts into the NDRC's project approval process. This "green accounting" (that incorporates subtracting the environmental damage done by specific projects) will yield more sustainable outcomes if such an integrated decision process values environmental damage at market prices. For example, suppose that a new highway is predicted to raise air pollution by eight units in a city. An environmental economist would ask that the damage (measured in yuan) from this extra eight units be subtracted from the project's benefits. To measure the damage caused by this pollution would require estimating how much individuals are willing to pay to not be exposed to this extra pollution and then adding this up across the affected population. While this is not an easy task, this valuation of nonmarket public goods would allow the central government to judge the net effects of a specific project before it is implemented, and this raises the likelihood that any major projects that are introduced are truly beneficial to society.

In the United States, environmental impact assessments are often required for major projects. In China the central government is creating new "rules of the game." Projects that may have serious environmental impacts now require the soliciting of public opinion, and the government claims that the assessment will experience a rigorous review process. For instance, the environmental assessment report on the Beijing–Shenyang high-speed rail project has been rejected three times in the past several years. But it still remains an open question whether a harsh environmental assessment of a specific project will lead the government to choose not to undertake it.

The MEP and the NDRC are the two main governmental institutions in charge of pollution mitigation, but they have different jurisdictions. The MEP is in charge of monitoring urban pollution, managing and publishing information, and treating pollution issues after they have occurred, and the NDRC is responsible for other environmental issues. Since the NDRC is a more powerful institution, its responsibilities also include formulating national political strategy and carbon emissions objectives. In contrast, the MEP's tasks are more focused on operations and how to achieve the NDRC's stated environmental objectives.

Besides the MEP, there are other ministries and administrations in the central government that are in charge of the environmental protection issues related to specific areas. Table 8.1 shows these institutions and their corresponding functions. Different ministries have overlapping jurisdiction, and they thus often fight over specific issues or take actions that impose costs on other agencies. This slows down the whole process and decreases the efficiency of environmental governance.

The MEP could build up its legitimacy and its influence by developing a reputation as an "honest broker" in evaluating the likely environmental impacts of different government projects. The agency is currently in charge of monitoring air and water quality; if it invests in expanding its monitoring network and distributes these data, it will have greater influence in

TABLE 8.1 Government institutions related to environmental protection

Government institutions	Functions related to environmental protection
National Development and Reform Commission	Environmental protection industry and regulation of its structure; climate change mitigation
Ministry of Water Resources	Protection of water resources
Forestry Administration	Forest conservation and ecology protection
State Oceanic Administration	Protection of the marine environment
Meteorological Administration	Weather changes
Ministry of Agriculture	Rural water and soil environmental protection
Ministry of Housing and Urban-Rural Development	Urban drinking water and waste
Ministry of Land and Resources	Soil and water conservation; soil improvement and protection
Ministry of Hygiene	Urban and rural drinking water health and safety

civil society because its benchmarks will be used to compare environmental progress both within and across China's cities. As we will discuss in the next section, information regulation has proven to be a highly effective tool at changing the behavior of both profit-maximizing firms and government officials.

Civil Society as a Regulatory Catalyst

Interest-group competition takes place every day in Washington, DC, and state capitals in the United States as environmentalist groups challenge polluters to pay for the damage they have caused. The traditional view is that the environmentalists face a David-versus-Goliath conflict in which they are no match for the lobbying resources available to private industries such as oil companies. There is an asymmetry in political power between the polluters and the victims (those exposed to pollution): the polluters are often small in number and well organized, while the victims of pollution can be large in number and face organizational costs that inhibit their ability to work together to lobby for environmental progress. In such a case, the tightly organized polluters will have the upper hand in the lobbying game because they can coordinate their efforts.

The US experience highlights a possible way for Chinese environmentalists to achieve the upper hand in their ongoing struggle for environmental protection. China's recent urban environmental woes have made the national news, and the central government has made no attempt to cover up or deny what is put forth in this news; a silver lining of environmental disasters is their work as regulatory catalysts, and unexpected environmental catastrophes capture wide public attention. Soon after five major such catastrophes—Three Mile Island, Love Canal, Bhopal, Chernobyl, and the Exxon Valdez oil spill—the US Congress voted on new risk regulation. In each of these cases leading newspapers such as the *New York Times* and the television news devoted ample attention to these issues. This saturation of the public media with stories about the issues lowers the cost for the average person to become educated about the issue, and this helps to overcome the classic "free rider" problem—the idea that no individual citizen has a strong enough incentive to take costly action to lobby the government for regulatory rule changes because he or she hopes that everyone else will take these actions, thus allowing him or her to enjoy the benefits of new regulations without bearing any personal cost.

In this sense catastrophes shock the public and put pressure on government officials to do something to lower the probability of future disasters. In the case of the United States, liberal/environmentalist congressmen are more likely to propose environmental bills shortly after environmental disasters take place, and the proposed bills are directly aimed at mitigating future catastrophe.[24]

The rise of civil society can play a productive role in helping citizens work together. Part of the credit for pollution mitigation progress in the United States is due to the rise of active environmental nongovernmental organizations (NGOs) that provide an independent source of information and help overcome free rider problems among the victims of pollution.

In the United States, nonprofits such as Robert F. Kennedy Jr.'s Riverkeepers launch lawsuits against polluting firms to stop companies from degrading the commons. China is now providing more room for citizens to organize themselves and express their environmental concerns, and a growing sector of (semi)independent NGOs addresses environmental issues. The number of Chinese environmental nongovernmental organizations (ENGOs) began growing in 2000 and experienced explosive growth within the last decade. According to Chinese Ministry of Civil Affairs statistics, in 2008 China had approximately 212,000 social groups, with 5,330 being of the environmental variety. Many Chinese ENGOs must simultaneously satisfy international donors and local government officials in order to survive.

According to Ruge Gao, three factors are fundamental to the birth, growth, survival, and influence of Chinese ENGOs: connections to the state administration, funding from corporate foundations, and the ability to promote public awareness through social and official media.[25] NGOs in China must be supervised and sponsored by state-run administrative agencies. This can become problematic if grassroots NGOs cannot find sponsors, especially if the NGOs' activity could be construed by the state as instigating political dissent. ENGOs with connections to government are relatively more likely to survive, simply from the benefit of being established as legitimately registered organizations through affiliations with government agencies. Through the powers of public media, ENGOs are capable of spreading environmental awareness and promoting their environmental and ecological goals, and this can help attract potentially interested volunteer groups and donors. The Institute of Public and Environmental Affairs (IPE), a registered nonprofit organization established by Ma Jun in Beijing in May 2006, has a different strategy to survive, however: its goal is to provide information, which is often obtained from official sources or self-reported by individuals, instead of confronting the government. This strategy is to some degree more effective in monitoring corporate environmental performance and facilitating public participation in environmental governance, especially compared with many extreme actions taken by environmental groups. International nongovernmental organizations (INGOs) and corporations have gained major public relations benefits by providing Chinese ENGOs with financial aid. INGOs have also invested large amounts of resources (economic and otherwise) in the development of Chinese NGOs; in fact, they are among the greatest supporters of Chinese NGOs through the financial grants and technical training they provide. Indeed, many Chinese ENGOs use foreign sources for most of their funding.

The various NGOs have begun to play a role in environmental protection in China, but the Chinese government passively responds to such public demands on an ad hoc basis, with little institutional commitment for engaging more public participation on environmental issues.[26]

Leading legal scholars focused on China's emerging environmental laws present a nuanced view of the current situation.[27] The good news is that in 2014 the government announced that it would raise the size of monetary fines that it imposes on polluters and require them to disclose more information about their actual emissions. The government now has the right to jail polluters, and some civil society organizations will have the increased ability to sue polluters to change their behavior. But these same scholars point out that there remain fundamental conflicts of interests at the local city level because the agencies that are empowered to reduce pollution have close political ties to the polluting companies. We will further discuss this point in chapter 9.

Academics have argued that civil society in China faces the challenge of figuring out "where the line is drawn" such that they will face a crackdown if they go too far in pursuing environmental progress. As Rachel E. Stern and Jonathan Hassid note, "Unsure of the limits of state tolerance, lawyers and journalists frequently self-censor, effectively controlling themselves. But self-censorship does not always mean total retreat from political concerns."[28] And Anthony J. Spires comments, "As long as grassroots NGOs stay small, make no calls for democratic reform, and allow officials to claim credit for their good works, their relationship with the authorities can be symbiotic and relatively smooth (if unequal)."[29]

Recent Cases of Environmental Activism in China

Wanxin Li, Jieyan Liu, and Duoduo Li provide two case studies of environmental activism in China.[30] The Xiamen PX chemical plant, with a total investment of 10.8 billion yuan, was one of the largest projects strongly supported by the Xiamen City government and approved by the NDRC in 2006. The second example is the Liu Li Tun garbage incineration power plant, which was planned in 2005 as an extension to an existing landfill that had been generating foul air and was close to reaching its full capacity. The plant would convert waste to energy, and was expected to be put in use before August 2008 when the Olympic Games started in Beijing. Both projects encountered opposition from different groups of environmental stakeholders.

The Xiamen PX Project

Chemical production has become a key industry in Fujian Province. The planned PX plant in Xiamen was only four kilometers from two university campuses and six kilometers from the city center, where 100,000 people live. In nations such as South Korea and Taiwan, such chemical production facilities are typically located more than seventy kilometers away from population centers. The benefit of a land-use policy that separates industrial activity from residential areas can be appreciated by noting the recent Tianjin disaster. In August 2015, Tianjin suffered a loss of over 160 lives as a chemical explosion took place in an industrial area close to a residential area.[31] The disaster highlights the danger posed to residential areas when risky industrial activity is placed nearby. This physical separation between chemical production and residential areas creates a type of "moat effect," limiting social costs imposed on downwind areas. Until recently the Chinese government had not implemented such urban planning principles. On June 1, 2007, fifteen thousand protestors marched through the city of Xiamen to protest against the PX plant. This demonstration was organized through text messages and demonstrated the ability of the people to spontaneously organize using information technology.

As Li and colleagues note, "Due to the public protest, on June 7 the Xiamen city government followed the suggestion of SEPA and announced a plan to conduct another environmental assessment before making a final decision on the PX project." A decision was reached to stop building the PX plant in Xiamen. On January 9, 2009, the MEP announced that the PX project would be relocated to Zhangzhou, Fujian Province.[32]

It is important to note that the chemical plant that was built in Zhangzhou has raised pollution levels there. In late July 2013, there was an explosion at the factory that generated wide concern.[33] This example suggests that local environmental activism may lead to domestic pollution "havens" as dirty activity moves to places where people complain less about exposure to pollution. Such locations are likely to be areas with poorer and less-educated households that are more concerned with economic development. In the United States, this dynamic helps to explain the environmental justice challenge that noxious facilities are often located in poorer areas with fewer homeowners. Areas with such demographics are less likely to mobilize to fight the opening of such a facility and are thus "rewarded" with greater exposure to pollution.

The Liu Li Tun Garbage Incineration Plant

In China there is a heated debate about what to do with the waste that citizens generate every day. The difficulty of producing a solution to the city's waste problem lies with the NIMBY, or not-in-my-backyard, mentality. People want their garbage to vanish, but they do not want it to be disposed of near where they live.[34] They also seek to pay a low price for disposing of their garbage.

As Li and colleagues note,

> The Liu Li Tun garbage incineration power plant was planned to be an extension of the Liu Li Tun landfill, built in 1996, which was expected to be closed down and turned into a park by the Haidian district government. The landfill is located in northwest Beijing, which is also the direction the wind blows from. It is adjacent to the Baiwang New Town community, which has hundreds of thousands of residents who have always complained about the stinky smell from the dump to universities such as Tsinghua and Beida to its south; and to hightech industrial parks both to its east and north. Furthermore, the Beijing-Miyun drinking water diversion canal is only 1.14 km to the south of the planned garbage incineration power plant. The project passed EIA in 2006 and construction was planned to start in March 2007; the power plant was to be put into use before the Olympic Games started in August 2008 in Beijing.[35]

Residents of nearby Baiwang New Town expressed concern about exposure to dioxin (a cancer-causing toxic compound) emitted from the burning of garbage; homeowners community expressed their concerns about stench and cancer on their community Internet bulletin board and petitioned SEPA to withdraw its approval for the power plant. The Development and Reform Commission (DRC) of the Beijing municipal government strongly supported the project and held a press conference to assure the public that the dioxin emitted would be within accepted safety levels.

But local residents did not believe the DRC's claim.[36] The Development and Reform Commission (DRC) of the Beijing municipal government strongly supported the 1.05 billion-yuan project and held a press conference on January 23, 2007, to assure the public that the dioxin emitted would be within accepted safety levels, but the local residents did not believe this claim.

On June 5, 2007, more than a thousand residents took to the streets to demonstrate in front of SEPA's headquarters in the Xicheng District. On January 20, 2011, the Haidian district government confirmed that they had dropped the plan to build the Liu Li Tun garbage incineration power plant. The new plant was planned instead for Su Jia Tuo, twenty kilometers from Liu Li Tun.[37]

One lesson from both the Liu Li Tun garbage incineration power plant project and the Xiamen PX project is that the central government's decision-making procedures should incorporate public opinion in its deliberations. The *China Daily* reported that the country needs a legal framework under which public participation becomes an indispensable part of policy making.[38]

As economists we would like to know why Su Jia Tuo was selected as the new site for the garbage incineration power plant. Does this area have a comparative advantage, perhaps because it is not heavily populated or because its location allows the emissions to broadly diffuse so that few are exposed to a toxic "hot spot"? If Su Jia Tuo was selected because it is a poor area whose citizens are not well informed, this would be another example of environmental activism leading to an increase in quality-of-life inequality as educated elites use their clout to protect themselves and to deflect a noxious facility to a poorer area. In the case of factory migration (which we discussed in chapter 3), such relocation of factories creates jobs in the poor regions where they end up. In the case of the siting of a garbage incineration plant, do local residents gain anything by living close to it? Recent research suggests that most urban protests are oriented toward enterprises that come from other regions or other countries, while less attention is paid to polluting enterprises run by locals.[39] One possible explanation for this fact is that residents are more likely to oppose pollution sources when they do not receive any benefits from the production.[40] There is no linear relationship between the intensity of pollution and the propensity of affected groups to protest. Whether citizens organize to oppose siting decisions depends on the specific circumstances of an individual case.[41]

An optimist might argue that if more people live in Baiwang New Town than in Su Jia Tuo, this relocation of the facility reduces total population exposure. These are the types of nitty-gritty details that must be examined in judging the environmental justice consequences of local NIMBYism. NGOs can play a key role in educating local communities about the actuarial risk they could be exposed to by such new facilities. Such information would allow for calm and logical negotiations so that the local communities are compensated for taking on new risks.[42]

The level of risk a new facility poses hinges on several engineering factors that the plant's management controls. If factories and noxious facilities anticipate that they will have to pay local compensation for any environmental damages they cause, this will incentivize them to invest in emissions control and other precautions to lower their emissions. Large companies that have many individual facilities and gain a good "green reputation" will face less public opposition when they try to open a new facility.

Environmental Litigation and the Chinese Courts

International research has documented that better-educated nations have stronger institutions and higher-quality governance.[43] Investment in an impartial judicial system is likely to be one major path through which a government may increase greater industrial compliance. If polluting firms anticipate that they face a serious probability of being found guilty of environmental malfeasance and face a significant fine, classic deterrence incentives will encourage such factories to take costly precautions that will clean the air and water.

For example, the Shanxi Yangcheng International Power Company which did not install desulfurization equipment, was fined 125 million yuan. If this company is aware that there is only a 5 percent chance that it will be convicted, the expected fine equals .05 x 125,000,000, which equals 6,250,000 yuan. Such a small fine provides little deterrence incentive and thus does not convince profit-maximizing firms to invest in costly precautions. Now let's contrast this with the same 125 million yuan but suppose there is a 90 percent chance that polluting firms will be convicted and fined. In this case, the dirty firm faces an expected fine of 112.5 million yuan, and this provides a much greater incentive for the same firm to invest in precautions. The net effect of credible fines is greater aggregate precaution and a safer and cleaner industrial China.

The institution necessary for bringing about this shift in expectations of accountability is a legal system with a transparent process for introducing objective evidence and rulings by judges and juries based on the evidence with no taint of political interference or whiff of corruption. When potential polluters anticipate that they will be held accountable for polluting, they will be more careful, and this will "green" their surroundings. In the case of US law, torts, liability suits, and class-action law suits have provided environmentalists with tools for incentivizing industrial firms to invest in precautions.

The number of successful environmental law cases is very small in China. The central government is making efforts to enforce the environmental laws and to encourage people to launch lawsuits against pollution accidents, but local courts are influenced by local governments.[44] The judges are appointed by the same level of government, not elected. In January 2015, China's highest court announced that it will reduce the costs faced by environmental groups for launching environmental lawsuits against companies or individuals that pollute.[45] The Supreme People's Court announced that it would give special status to help launch public interest litigation; this appears to be a very promising initiative. In the same month, an eastern Chinese court ordered six companies to pay a combined total of over 164 million yuan (US$26 million) in fines for discharging acid into two waterways. The fines represent the highest penalty in Chinese environmental public interest litigation thus far.[46] Yet whether this incentive will lead to more lawsuits, and whether these lawsuits help hold polluters accountable, are important questions.

The Reduced Information Costs of Monitoring and Discussing Pollution Challenges

While the Chinese government controls the leading newspapers, the rise of Internet media and microblogs have allowed individuals to express their concerns and displeasures with quality-of-life outcomes in China. The microblog as a nascent web application emerged in 2009 in China, and its usage has surged since then, with 250 million users by the end of 2011. The national government has also started to use Internet media strategically as a tool to bring about accountability and transparency on environmental issues. Its role has been highlighted in the Xiamen PX protest of June 2007, the Dalian PX protest of August 2011, and the Beijing particulate matter debate in October 2011.

The media devotes ample coverage of such salient issues as particulate matter (measured as $PM_{2.5}$) air pollution but pays much less attention to less-visible pollution such as heavy metal emissions. Media reports also tend to focus more on specific events (e.g., the PX protects) than they do on pollution problems that develop gradually. As a result, local government agencies increasingly feel pressure to immediately respond to media reports while paying less attention to less salient issues.[47]

Improvements in information technology have reduced the cost of information acquisition for both the central government and Chinese urbanites. Access to this information has allowed local people to be better informed about the specific pollution challenges they face. This news allows them to overcome potential free rider issues and to unite to express their concerns and displeasure with current urban quality of life. Since social stability is an important target when the state evaluates local officials, those officials are keen to address their people's demand for a cleaner environment.

Recent research set in India highlights the powerful role that the media can play in setting environmental priorities. Esther Duflo, Michael Greenstone, and Rema Hanna document that air quality regulation has been more effective than water pollution regulation in India.[48] They studied the content of the *Times of India*, a leading newspaper, and found that there was much more media coverage of air pollution than water pollution. Such citizen interest, both a cause and an effect of the media coverage, persuaded India's Supreme Court to become active in the implementation and enforcement of the air regulations.[49]

Continuing Challenges

Local officials in China are often better informed than the central government or the citizenry. Mayors have private information about their cities' environmental performance and may cover up environmental challenges to minimize concerns from the central government. China's central government has built channels to monitor local government officials; the channels traditionally used to collect information have been government audits and citizen petitions, but their effectiveness has been questioned.

There have been many reports in the Western press of people being intercepted on their way to file petitions; some of them end up in "black jails."[50]

Recent academic studies highlight the continuing challenges the central government faces. One study has documented that industrial deaths at factories that had social connections with local government officials were higher than at factories without such connections.[51] One possible explanation for this fact is that local Chinese officials take good care of their friends and do not enforce the laws evenly across such factories. These "connected" factories have a cost advantage in production because safety precautions are costly, but they are more likely to suffer accidents. This research suggests that in cities where industrial plants are major polluters, local officials may pursue their own agenda in determining how much regulation these factories must engage in even if the central government seeks to hold these plants accountable for their pollution. We will return to this point in chapter 9.

Another study has used data from air pollution monitoring stations to document that local officials are strategically manipulating data that is reported back to the central government. The intuition behind this research strategy is that there are discrete cutoffs for the air pollution index (API) score that dictate whether a day counts as *clean* versus *hazardous*. The researchers plotting these data note an asymmetry: high-pollution days are much more likely to be recorded at just below the API threshold. This means that the city's leaders are recording clean days too often relative to the actual monitoring of local pollution. Again, this serves the strategic interests of the local mayors as they announce that their city is clean, but it impedes the efforts of the central government to classify cities that are making environmental progress versus those that continue to suffer high levels of pollution.[52] Both of these studies suggest that local officials continue to have private information and discretion and can take actions to protect their friends and send signals to the central government that they are trying to comply with the rules when in fact they are being selective about such compliance.

If the central government understands that it faces these challenges, what can it do to respond? In the case of local air quality, the government could require that the local authorities provide the full daily data on ambient pollution rather than providing such data by discrete cutoffs such as the total count of clean versus dirty days per year. Let's suppose that a day is classified as very dirty if the $PM_{2.5}$ reading exceeds 100 and is classified as clean if the $PM_{2.5}$ reading is less than 100. Suppose that 122 days in the year 2014 receive a $PM_{2.5}$ reading of 80, 122 days receive a reading of 99, and 122 days receive a reading of 101. In this case, if this city reported its data accurately to the central government it would report that 66 percent of the time its citizens enjoyed blue skies. Now suppose that it could shade down the $PM_{2.5}$ readings of 101 to 99.9999. This is a very small quantitative shading, but now the city could report that its citizens enjoyed blue skies every day of the year! The simple solution to this data-reporting issue would be to require the city to report basic statistics for its overall pollution distribution each year such as the mean, median, and the 99th percentile of the distribution.

If the central government worries that local factories are collaborating with local officials and not truthfully reporting polluting activity, the government could sponsor random audits of factories and severely punish local officials if violations are spotted. In the case of air pollution monitoring, it is getting less expensive to monitor air quality, and this should reduce information asymmetries. The central government could introduce its own system of monitoring stations and compare its readings to local ones.

A third challenge that the central government faces is that many of its environmental targets intended to evaluate local officials' efforts on green criteria are not closely linked to environmental outcomes that have significant impacts on the public's health and quality of life. Instead, they are linked to accounting indicators such as energy intensity and environmental infrastructure investment. Credit for pollution reduction might be granted, for example, for the construction of a waste gas treatment plant or installation of pollution control technology in a power plant. Local officials are therefore incentivized to invest in environmental infrastructure and pollution control technology. With insufficient monitoring, there has been much less emphasis on whether these investments are operated properly such that they actually reduce pollution. It has been reported that factories have adjusted pollution control equipment to report false data, treatment plants have been left idle, local governments forced emergency shutdowns of electricity to local public services to meet energy efficiency targets, and so on. But Chinese citizens do not care that China has installed an unprecedented number of flue gas desulfurization units in power plants; they care about clean air. This suggests that the central government must devote more attention to how it designs its evaluation criteria. If it continues to focus on inputs such as installed desulfurization units, it must have a clear understanding of how such investments translate into improved environmental outcomes.

The Legacy of Communism

When choosing national resource pricing, the central government faces the trade-off between increasing consumer purchasing power and protecting the environment. Energy policies such as subsidized winter heating north of the Huai River were intended to improve quality of life but have unintended consequences for the environment.[53] Recent research has documented that providing highly subsidized winter heating for the five hundred million Chinese who live north of the Huai River caused an aggregate loss of 2.5 billion years of life expectancy.[54]

Residential electricity and water use are also underpriced in Chinese cities. Fuel prices were once highly subsidized in China, and these low prices encouraged people to commute long distances from low-density suburban areas. Fuel prices have been rising quickly in recent years, but no formal gas tax is collected. By setting resource prices low, the state raises consumer purchasing power and lowers firms' costs of production, which aids growth, but these policies exacerbate externalities through encouraging excess consumption. The central government would be better able to tackle urban pollution externalities by using incentives such as removing price ceilings for gasoline and winter heating. As China's economic growth continues and the count of people living below the poverty line decreases, the government may be more willing to allow prices to rise as its concern for "price gouging" the poor would decline.

Now that we have analyzed the incentives of the central government in determining how it prioritizes blue skies, we turn to priorities of China's different mayors and the trade-offs that they face as they lead China's diverse set of cities. Such cities differ with respect to their current environmental challenges due to their geography and their size, industrial structure, and population demographics. Mayors differ with respect to their age, education, and ambitions. This matching of different mayors to different cities introduces a wide range of responses to the ongoing set of environmental challenges that China faces.

CHAPTER NINE

Will Local Governments
Create Green Cities?

China's former president Hu Jintao graduated from Tsinghua University fifty years ago, and its current president, Xi Jinping, obtained his bachelor and doctorate degrees from Tsinghua. Many Tsinghua graduates, especially those who have earned doctoral degrees, have entered various levels in China's government hierarchy and later often become local or national leaders.

A dozen of coauthor Siqi Zheng's Tsinghua friends now hold leading positions in the key departments in small- and medium-size cities' governments across China. During each year's University anniversary days, many of them return to campus and gather together to exchange ideas. From those gatherings Siqi has learned about urban quality of life in second- and third-tier cities. In recent gatherings a major topic has been the environmental challenges China faces. We now share the stories of some Chinese mayors.[1]

What Policy Goals Do Mayors Prioritize?

Local officials all agree that, before 2000, economic growth was the first priority for both the central and local governments; at that time environmental issues were barely discussed. For a long time, China's central government focused on economic growth, with an emphasis on gross domestic product (GDP) as the key evaluation criteria of local officials' performance. In those "good old days," mayors in Chinese cities had clear marching orders. If a city enjoyed rapid economic growth and if the city launched large scale infrastructure projects, such as building new highways and large industrial parks that boosted its economy, its mayor would be much more likely to be promoted. Ambitious mayors who sought to be promoted to higher political levels (such as to the rank of provincial governor or central government official) had strong incentives to play by these rules.

Mr. Shi is now the mayor of a county-level coastal city in northeast Zhejiang Province, just outside of Shanghai. Before he became the mayor, he served as the director of the Merchants Department in city government; this department's main task is to attract firms to the city. He notes that inexpensive land and subsidized energy were the major policy instruments used by the city to attract manufacturing firms. Such subsidies posed a trade-off for his city; in the short run, the city lost tax revenue, but in the long run the city government expected to collect greater tax revenues from such firms. Mr. Shi's predecessor told us that during his time as director his office preferred luring export-oriented manufacturing firms; it was not concerned that such firms would contribute to local industrial pollution challenges.

In the past, city mayors intentionally sacrificed environmental quality to achieve their goals of economic growth and high tax revenue. Local governments needed the taxation revenue and economic output from polluting industries, but they had no incentive to address the pollution costs that such economic activity imposed on areas outside their political boundaries. As we discussed in chapter 2, many factories are located along rivers in China so that the area acquires jobs and economic activity while the pollution damage to the river is borne by people downstream.

Today cities differ with respect to specific attributes that will influence their mayor's willingness to enforce environmental regulations. Consider a city whose economy does not specialize in manufacturing and whose health expenditure is a large share of the budget. This city will be more likely to pursue the green agenda, and the mayor will recognize that few citizens will experience losing their jobs because of tighter industrial regulation. Given that pollution has a direct impact on the population's health, if the city is incurring large health care costs, a mayor will seek to lower pollution to improve the population's health and thus to reduce its reliance on local health care.[2]

Mr. Shi recently became the mayor, and he has imposed some stricter environmental regulations. For instance, he requested that all firms emitting sulfur dioxide must install desulfurization devices and install monitors detecting whether those devices are in use. He also installed monitors both upstream and downstream from the industrial firms. While such factories have private information about their daily activities, if the downstream water monitoring station's pollution level is much higher than the upstream monitoring station then this suggests that the factory is polluting the water.

Mayor Shi's monitoring strategy mirrors a standard technique in academic economics focused on measuring value added by an instructor.[3] Consider a group of students who take a standardized test at the start of third, fourth, and fifth grades. If Ms. Smith is one of many fourth grade teachers and a researcher observes that the average increase in the test score between the fourth and fifth grades is greater for her students than the average overall increase in test scores between fourth- and fifth-grade children at the same school, the researcher will conclude that Ms. Smith ranks highly as a value-added teacher. While researchers cannot observe any individual teacher's effort in the classroom, her students have enjoyed a greater improvement on a standardized metric than other children of the same age at the same school.

Mayor Shi's monitoring approach establishes the *pollution added* by the major industrial plants by sampling the water quality before and after the water passes the plant. The difference in pollution levels reflects the industrial plant's contribution to pollution.

Zhejiang Province uses this approach to evaluate the emissions from its firms. Mayor Shi notes, "The environmental protection agencies of Zhejiang Province regularly sample the water at both upstream and downstream points inside and adjacent to my city's boundary to see if the water quality gets worse after the river flows through my city. This data is directly reported to the provincial government without letting me know." Another effective evaluation tool the Zhejiang provincial government uses is the collection of people's opinions through random phone calls. "The staff working in the provincial government's environmental protection agencies randomly call the residents in my city and ask them whether they feel their quality of life is good, and whether they like their current mayor," Mayor Shi explains. "This phone survey will be used when evaluating my performance for the year. More important, if you know your people do not like you because their environment is polluted when you are the mayor here, you will definitely feel guilty!"

Some Chinese cities have changed course and are no longer are seeking to lure heavy industry to locate within their borders. Mr. Shi took over the director's position in 2003. At that time his boss, the mayor, began emphasizing that their city did not want to attract dirty firms. The city stopped offering polluting industry generous recruitment incentives and directed such firms to contact other small and underdeveloped cities' governments.

Mr. Shi understands the importance of environmental quality as a key determinant of quality of life for the residents in his city, which is at an advanced stage of economic development and is quite attractive for companies. As a consequence, he has the ability to choose which enterprises to accept without worrying that refusing a specific firm can significantly injure the local economy. He knows that if his city achieves greater GDP growth, fiscal revenue growth, and foreign direct investment growth that this will improve his chances for promotion. The mayor recognizes that environmental targets must be achieved because the higher levels of government place great emphasis on not exceeding these emissions levels. A common practice for city mayors is thus to only recruit those firms that meet baseline environmental targets; mayors will then pick the most productive ones from among those firms. The residents' income is gradually increasing, so they demand a higher environmental quality. Such residents applaud their local government's efforts to require that firms increase their environmental control investments.

Environmental Leadership

While it is common sense that leaders matter in determining a city's quality of life, academics have had trouble testing this claim. For any given city currently enjoying environmental progress, it is difficult to judge whether policies are the cause of the progress or whether some other factor such as the business cycle or international trade is the true cause. The old point made in every statistics class that "correlation does not imply causation" merits repeating. If a city enjoys environmental progress during a specific mayor's leadership, do we give him the credit, or were there preexisting trends such as deindustrialization taking place so that the city would have enjoyed environmental progress regardless of who was mayor?

In our own research we have posited that urban leadership is key for causing environmental progress in China's cities. As the Chinese saying goes, "a new official always brings in new rules." New leaders have their own ideas on how to develop and manage their jurisdictions.

In our research studying promotion prospects, we found that in cities with a younger and more educated mayor, greater environmental progress took place. This suggests that as a new generation of environmentally conscious mayors take office they may pursue a local growth strategy based on the cleaner practices found in, for example, the service industry, high technology, and tourism. Our statistical analysis shows that if a new mayor has a higher educational attainment than the former mayor, the city enjoys a significant improvement in air quality. A one-year increase in the mayor's years of schooling is associated with a 2 percent decrease of the average value of air pollution as measured by concentrations of particulate matter up to ten micrometers in size (PM_{10}).[4] While this effect may appear to be small, it is important to note that this is the size that we would expect for the time period when we estimated it (the years 2004–9). The important point from our perspective is that this statistically significant effect is found in the historical data, and we expect that the magnitude of this effect will grow in the future. In our peer-reviewed research, we have documented that cities with more educated mayors reach the environmental Kuznets curve turning point at lower levels of per capita income. This means that leaders can influence the relationship between local economic growth and pollution levels.

Some local officials do not necessarily want to get promoted (especially if they have cozy deals with local businesses or personal reasons to stay local), or know that their odds of being promoted are not that great (especially if their educational credentials are mediocre). Including environmental targets in the promotion criteria would not create a strong incentive for this group of local officials to protect the environment. This discussion highlights that the central government's rules will not incentivize all local officials to "go green." This means that further empirical research is needed to identify how many officials, and in which cities, fall into these categories. Our own empirical work suggests that many mayors are responding to the new incentives. For example, we find evidence of greater environmental progress in cities with younger mayors. These mayors have a longer political time horizon, and this finding supports the claim that older mayors who have a shorter political horizon may be less likely to pursue a blue skies agenda. But the central government can easily observe each mayor's age and engage in extra monitoring in cities whose leaders' demographic profile suggests that they will shirk on their environmental responsibilities.

Local Governments' Incentives for Protecting the Environment

When we ask Mayor Shi why he is pursuing the "green development" path, he explains that the logic is quite clear. His bosses are the governors in the upper-level prefectural city government and Zhejiang Province government. When those bosses evaluate his performance, they will consider his efforts in protecting the environment. He is quite aware that the central government now includes green targets in the performance evaluation and promotion criteria of local government leaders to encourage such leaders to pursue a sustainable development strategy. Mayor Shi told us that the targets that they pay most attention to are the GDP growth rate, fiscal income, industrial value added, exports, and foreign direct investment (FDI). These targets are not binding ones with veto power (*yipiao foujue*). Environmental and energy consumption targets are veto-power targets and he has to fulfill them; otherwise he cannot pass the end-of-year evaluation.

Local officials like Mayor Shi are well aware that the central and upper-level governments are judging their performance based on a set of criteria that increasingly place weight on environmental criteria. An open question is whether the data reported to Beijing can be manipulated to convey a rosy picture of a city's environmental performance (even when the truth is that local quality of life is terrible) or whether these criteria correctly convey the true state of environmental quality. Only in the latter case will this process identify cities that are making legitimate environmental progress.

If the central government is aware that local mayors have an incentive to exaggerate their environmental progress to Beijing, the central government has a simple counterstrategy. In the past, the central government obtained most environmental information from the reports submitted by local environmental protection agencies. Such reports suffer from an obvious conflict of interest. As the price of monitoring environmental quality using remote sensing and random inspections declines, the central government can conduct more audits. If the government announces that it will heavily punish local officials who report environmental excellence while independent audits suggest a lack of environmental progress, this would provide a strong incentive for local mayors to truthfully report the current state of their city's environmental situation. This example highlights that if the central government "knows that it does not know" certain key information, it can design an incentive system that encourages the mayors to reveal their private information and to devote more effort to actually improving environmental conditions.

Since 2000, China's Ministry of Environmental Protection (MEP) has been directly monitoring daily air and water quality in roughly one hundred cities, and it posts those official pollution data sets collected from ambient pollution monitors on its website. Since 2007, the central government has begun to release annual data listing the emissions of key monitored polluting firms. The emissions from those firms are directly monitored by the MEP.

Such information revelation solves a pollution challenge in that urban leaders have inside information over the key local firms' pollution emissions into the air, groundwater, and underground water, and they have inside information about their own environmental regulation enforcement efforts in their cities. In the absence of randomized audits, local government officials have weaker incentives to enforce such regulations in their own jurisdictions.[5] In the past the local government had a news monopoly over the environmental disasters in its jurisdiction. Those disasters might not been known by the public or even by the central government, and in the absence of such information, industrial malfeasance would be more likely to take place as firms would underinvest in precautions.

While we are optimistic that local leaders have increased incentives to pursue green objectives, there are other features of the Chinese system that discourage prioritizing sustainability goals. Implementation of environmental targets is made harder through the institutionalized cadre rotation system that switches leading officials to new positions or localities every three to four years.[6] This incentivizes local leaders to prioritize short-term over long-term gains. For instance, they may extract rents from local polluting industries rather than close down those dirty firms. In the academic corporate finance literature, researchers have often advocated that CEOs be paid with company stock options to incentivize them to take actions that boost the long-term growth of the company;[7] companies such as Amazon are known to pay senior employees with stock options for just this reason. The urban analogue would be to pay mayors with future plots of the city's land. Such land will be more valuable in the future if the long-run quality of life and economic base remain strong. We know of no city around the world that has pursued this compensation scheme to incentivize local leaders to adopt a long term perspective.

Environmental Challenges Posed by China's Multilayered Governance Structure

Mayor Shi has also voiced the concern that he does not have enough power to fully control the pollution in his city. A large state-owned power plant company has a branch plant located in the suburban area of his city. Mayor Shi asked this branch plant to install desulfurizers and report the daily operating condition of these facilities to him, but the plant did not cooperate. In China's political hierarchy, the senior managers of this state-owned enterprise (SOE) hold a higher position than Mayor Shi, so they ignored his request. Since this power plant has brought in a large amount of tax revenue and plenty of job opportunities to his city, Mayor Shi has to be careful not to offend the managers. This phenomenon plays out throughout China. There is a certain irony that the SOEs, rather than purely for-profit firms, could become the major sources of industrial pollution because their leaders recognize that they are often above the local law. Many small- and medium-size cities are faced with major SOEs that deliberately violate environmental regulations. The challenge is that these large SOEs produce economic output and generate jobs that are essential to local governments.

An active research topic in academic macroeconomics focuses on productivity differentials of private enterprises versus SOEs in China. One study estimates that China's overall productivity could be much higher if the SOEs' share of the economy shrank.[8] Such SOEs do not have strong enough incentives to be efficient firms because they can access central government funds that subsidize their inefficient operations. Lower-middle-class workers would counter that this "inefficiency" means jobs for people like them even if the value of the output they create does not exceed the wage they are paid.

Given our focus on China's green cities, the relevant issue here is that macrostudies of productivity ignore social externalities such as pollution. If Mayor Shi is correct in noting that the SOEs contribute to his city's pollution problem and thus exacerbate social externalities, macroresearchers who measure SOE productivity are *underestimating* the productivity impact of SOEs on the Chinese economy because they do not consider such firms' pollution impacts. Put simply, if the central government goes ahead and engages in SOE privatization, this could simultaneously raise the productivity and lower pollution in many Chinese cities. Such a privatization would not be a free lunch; displaced workers would face the transition challenge of finding new jobs without suffering a large wage cut. A US-based labor economics literature has found that manufacturing workers suffer large wages cuts when they lose their industrial jobs and transition into working in the service sector.[9]

While local leaders have increased incentives to address quality of life within their jurisdictions, they have weaker incentives to recognize the regional impacts of their policies. In the decentralized Chinese pollution enforcement system, environmental improvement targets are allocated from the central government to provinces, and then from provinces to different departments within administrative boundaries of a province, municipality, or county. These departments are responsible for just the area within their corresponding administrative boundaries. But lakes, rivers, and wetlands are ecosystems that should be managed as single entities rather than parceled out to different administrative units.[10] An example is the recent $PM_{2.5}$ pollution challenge in Beijing, whose air pollution results partly from coal burning in neighboring provinces—especially Hebei Province, which burns two hundred million tons of coal every year. Without interprovincial joint efforts in the greater Beijing area, the city's municipal government can do little to stem local pollution. There have been some efforts to solve the environmental governance problem regarding such cross-boundary issues. For example, in 2013 the Environmental Protection Bureaus in Anhui, Jiangsu, Shanghai, and Zhejiang Provinces initiated the Yangtze River Delta Transboundary Environmental Pollution Emergency Response Plan. Under this new cooperative measure, provincial governments committed themselves to sharing environmental protection resources and jointly investigating and addressing cross-border pollution incidents. While it is too early to tell whether signing such an agreement will actually achieve measurable improvements, this does signal the awareness among local leaders of the need to improve cooperation across provincial borders. In chapter 2 we discussed Hong Kong's attempts to incentivize neighboring polluters to internalize the social costs of their production on Hong Kong. We predict that similar resource transfers will take place as those who are suffering from the damage offer resources and technologies to try to mitigate the challenge.

Mayors Differ with Respect to Each City's "Golden Goose"

Mr. Han, the vice mayor of a county-level city in Hubei Province, also graduated from Tsinghua University. He is younger than both Siqi and Mayor Shi. During a dinner when we all sat together, he disagreed with Mr. Shi's "aggressive" greenness strategy. Mayor Han's city is much poorer than Mayor Shi's—the former's per capita income is just about one-half of the latter. There are two large paper mills and many small ones located in Mayor Han's city, and they are the city's main tax contributors and employers, but they are also major emitters of wastewater into local rivers. "Your city is rich, so you can move away dirty firms, but I have to balance among many things," Mayor Han said to Mayor Shi. "People in my city need jobs and want to see that their income is growing. I also want to achieve those goals to show I am a capable mayor." Mayor Han said he also asked the paper mills to install wastewater treatment facilities, and the water pollution is under control, but he emphasized that he would not set a very strict water quality requirement for those firms and does not plan to monitor whether the wastewater treatment facilities are 100 percent in use. The Environmental Protection Bureau in his city has not received any additional staff in the past ten years despite rising environmental regulations and new responsibilities, and the current staff lacks proper training to properly monitor those facilities. "Everything needs a process," Mayor Han said. "My boss [the provincial governor] understands my situation well. He knows it is hard to kill two birds [economic growth and a clean environment] with one stone in my city for now. Of course, I believe, ten years later, if I am still the mayor here, I can do the same thing as Mr. Shi does today."

The pollution haven hypothesis posits that for-profit firms will seek out geographic areas where regulation is less likely to be enforced. More industrial accidents are likely to take place in geographical areas with lax regulation. This is more likely to happen in China, and in less-developed countries where the informal sector is an important part of the economy. Illegal rare-earth mining in China is a prime example. China mines 99 percent of the global supply of heavy rare earths, with legal, state-owned mines accounting for only a small portion of the nation's output. Mayors are more likely to give mining permits to those mines in far suburban areas so that they can obtain the revenue from selling rare earth but also keep mining activity far away from dense population centers. There is an enormous black market for this output. China's national and provincial governments are working hard to crack down on the illegal mines to reduce their severe environmental impacts (to which local authorities have long turned a blind eye).[11]

Mr. Fang is the vice director of the Development and Reform Commission of a coastal province, and he told us an interesting story. Cities in his province exhibit different attitudes toward environmental protection; whether local mayors pursue national green goals hinges on what they perceive to be a "golden goose" for their political career. Mr. Fang told us that the mayors in inland cities are not that active in protecting the environment, but the two tourist destinations in his province are keen to pass stringent environmental regulations, since the mayors understand that blue skies and clean water are crucial for attracting tourists and supporting the local economy. As China builds up its air travel network, wealthier people have a larger set of destinations (including international ones) to choose from. In order to be competitive against alternative tourist destinations, China's tourism cities have strong incentives to maintain beauty. An example is China's Sanya, in Hainan Province, which calls itself China's Hawaii.

In Mr. Fang's province, coastal cities are generally wealthier than inland cities, and the mayors in the coastal cities place more emphasis on green issues because they face less pressure to deliver GDP growth. These cities understand that their economic niche is high technology and skilled industries, so they pass environmental regulations seeking to improve urban quality of life because this helps attract high-skilled human capital workers.

As the *hukou* internal passport system is dismantled, Chinese urbanites can increasingly "vote with their feet" and move to the city that best suits their needs. Those cities that offer gray skies and low quality of life are likely to experience a brain drain. In this sense, the threat of mobility and "voting with one's feet" is a key way to potentially discipline urban politicians if they do not pursue the promise of blue skies. We recognize that this potential brain drain will not scare a mayor who views his city's future as tied to industrial production. Beijing, in part because of its historical importance and because it is a capital city, has a unique ability to retain talent despite its high levels of pollution. There are other Chinese cities, such as Xiamen, that are positioning themselves as high-amenity, high-quality-of-life cities. We expect that such cities will increasingly attract affluent, educated people, much like Los Angeles and Miami do in the United States.

Those cities that adopt aggressive environmental regulations are likely to lose some manufacturing jobs. A general conclusion from US academic research is that polluting firms (especially those who are footloose because their production is not tied to some key input that is heavy to ship across space) seek out areas featuring less regulation.[12] If China's wealthier cities impose more stringent vehicle and industrial regulation, this will displace dirty industries to domestic pollution havens; many energy- and labor-intensive firms in China's coastal cities have relocated to the central and western regions. Mr. Fang explained that, in his province, if those firms want to leave the coastal cities, he and his colleagues encourage them to move to the inland cities within the province by giving them subsidies and tax deduction incentives. As a result, the firms still contribute to his province's economic output and fiscal income yet at the same time help the relatively poorer inland cities to develop. Mr. Fang noted that other provinces also use this strategy. Guangdong Province is a good example; it implemented the "ridding the cage of old birds [polluting firms] in favor of new ones [clean and high-skilled firms]" policy in the Pearl River delta in 2007. The key piece of this policy is subsidizing polluting firms in the delta to relocate in the northern part of the province, and Jiangsu Province is relocating those firms to the north Jiangsu (Subei) area. Provincial governments pursue this strategy to simultaneously "green" the big city and spread income (and new industrial plants) to the poor underperforming areas within the same region.

Mr. Fang also told us about some unintended consequences introduced by well-meaning central government policies. Including energy saving and pollution reduction targets in local officials' evaluation criteria sometimes leads local officials to engage in strategic behavior. By the end of 2009, national energy intensity levels had been reduced by merely 14.4 percent, far short of the expected eleventh five-year plan target of 20 percent. As a response, many subnational governments undertook last-minute measures to meet their energy intensity goals. For example, one local government cut off electricity to homes and rural villages, even to the extent that one hospital was forced to close once every four days, and another city implemented a "work five, stop ten" power rationing practice for large enterprises, which was equivalent to working only ten days per month; as a result, local employees could earn only one-third of their usual wages. Some enterprises facing electricity rationing switched to diesel-operated generators, and this actually increased particulate pollution. These examples highlight that the central government could do a better job anticipating the strategic responses from local governments who seek to comply (in the cheapest way possible) with the new rules. In some cases, a well-meaning goal can backfire.

Do Tighter Environmental Regulations
Reduce Industrial Productivity?

In the early 1970s some economists posited that the observed industrial productivity slowdown in the United States was caused by the growth of environmental regulation. This conjecture was based on the fact that US productivity was slowing at the same time that the Environmental Protection Agency had been launched and its regulatory authority had increased. Whether this correlation represented a causal relationship continues to be debated.

If this conjecture is true, an unintended consequence of China's pursuit of green goals would be to slow down its industrial productivity growth. For cities where industry plays an important, ongoing role in the economy, can industry be regulated without sharply increasing its costs? If such regulation does impose costs on industry, will they respond by shutting down factories and moving to poorer nations where they will not face regulation? Will they fire workers and reduce production, or will they seek to raise prices to end consumers? These same issues arise all over the world when tighten environmental regulations are debated, and they boil down to the question of who pays for environmental regulation; workers, consumers, or the capitalists who own the firms.

The answer partially hinges on whether the firm works within a perfectly competitive market or if it has some monopoly power. In the second case, the firm can raise prices as its costs increase, and end consumers will be likely to pay a higher price for the product; some workers will lose their jobs as the firm produces less. For goods where China's firms work within a perfectly competitive world market, consumers will always have plenty of alternatives and they will substitute other goods for Chinese goods if the producers try to raise prices. In this case, Chinese workers and the owners of the firm would bear the costs of the binding environmental regulation.

Today there is an emerging consensus that there are significant differences in the ability of different firms in the same industry to adapt to new regulatory conditions. Due to differences in the age of the factories and the quality of their managers, some firms will have more trouble than others in complying with new emissions reductions requirements.

Some academics such as Harvard Business School's Michael Porter have even gone as far as claiming that environmental regulation has negative costs! The controversial Porter hypothesis posits that firms suffer from a status quo bias such that their leadership is often "asleep at the wheel."[13] In such cases regulation acts as a wake-up call and encourages firms to take a second look at their operations; this helps to eliminate waste and lowers costs. As China's cities ramp up their environmental regulations, the Porter hypothesis will be tested. A future researcher could investigate whether the productivity of regulated firms and their stock market performance improves relative to similar firms that continued to produce in areas with weaker regulations.

A Story from an Underdeveloped City in Western China

Siqi's doctoral student Cong visited a small town in the west of Guizhou Province in the summer of 2013. This town has more than ten billion tons of high-quality anthracite coal as a natural resource; and the central government has recently begun allowing some coal mining to take place to generate some income for the local economy. Many mining firms entered the small town, and among them was W, a new state-owned coal mining firm. In 2008, W began constructing two mining pits that will be operational by 2016. Cong talked with some managers of this firm, who explained that it was very difficult to operate: they had to keep good relations with local government officials and local residents, and land acquisition was a problem. Many residents had been living or farming there before their firm was given the land from the local government, and the firm was now responsible for relocating those residents to other places and to compensate them, then transform the farmland into industrial land. The managers said this process did not go well until local officials helped them.

This new coal mining activity poses several environmental challenges beyond destroying the land surface. Coal exploration would pollute underground water. Unlike the challenge of air pollution, such underground pollution might occur without local residents being aware of it. Mr. Li is a local official who works in the Bureau of Coal Mining Administration in this town; he noted that more than 70 percent of the town's revenue is

Figure 9.1 A coal mine in a town in Guizhou Province
Photo credit: Cong Sun, August 2013.

from coal mining industry (see fig. 9.1). The town's residents understand that they face the dilemma of environmental protection versus attracting industrial development.

Different mining firms impose different social costs on the town. Larger firms have the scale and the profits to pay for the extra costs of a modern pollution treatment plant, and this will reduce their pollution per ton of coal processed, but the small mining firms cannot build such facilities; these latter firms will have lower unit costs but produce more pollution per unit of coal. This example suggests that larger entities may have cleaner production techniques because they will invest in emissions control equipment, and such firms are likely to anticipate that they will face greater regulatory attention because of their size. If smaller firms and large SOEs face less regulatory attention (for the reasons we have sketched), China's cities could enjoy environmental gains by simultaneously encouraging horizontal mergers of smaller firms and the privatization of the SOEs. This process would yield an industrial organization featuring fewer, larger firms that are private companies independent of government. These companies would be large enough to be "on the radar" of government and would have the financing and the human capital to invest in pollution abatement.

Encouragement from the Urban Public

The urban public represents an important constituency, putting pressure on mayors to "green" their cities through the central government. For example, the central government can push local governments to enforce unpopular measures to alleviate public protests. In the past, the Chinese public faced greater information costs concerning the environmental challenges their cities faced. The state and local governments monopolized the media (newspapers, radio, and television), and when a one-party state controls information it may systematically choose to release information that helps it to achieve its political goals but may suppress negative information.[14]

As the new urban cohorts in China become wealthier and more educated, they have a greater willingness to pay to avoid risk; their demand for information and political accountability is thus likely to rise, and the rise in demand for environmental information creates incentives for the media to cover such stories. In recent years, the Chinese media has devoted much greater attention to environmental issues. For instance, the number of Google news items on this topic for 2011 was 240 times that of ten years earlier. Improvements in remote sensing and less-expensive pollution monitors have allowed external research teams to measure and distribute information about China's pollution levels.[15]

Research from Brazil, India, Indonesia, and the United States highlights the power of the media and information disclosure to encourage government officials to supply public goods.[16] The studies have documented that local officials do a better job delivering higher-quality public services when they are aware that citizens are informed about their performance; citizens can be informed either because they are well educated and are reading local newspapers or because of reductions in the cost of acquiring information. An example of the second case is providing citizens with nonpartisan report cards on the objective performance of individual government leaders; such trusted information allows citizens to update their beliefs about the quality of their officials and this allows them to make more informed choices during elections, something that is effective in these countries because they are democracies. Anticipating this dynamic incentivizes local officials to do a better job in providing public services.

Will rising information transparency in Chinese cities play a similar role? The recent upsurge of environmental incidents provides some clues. In these mass incidents, the media coverage facilitated public coordination and helped people to overcome the obstacles often associated with organizing protest against pollution accidents. A milestone event is the Jilin chemical plant explosion that polluted the Songhua River in northeast China in 2005.[17] This major water pollution event affected downstream provinces and, later, Russia. Many media institutions started to report details on this accident and criticized the local government's slow response; the timely reports were welcomed by the public. Since then the media has been active in reporting large environmental pollution incidents; representative examples include the Xiamen PX protest in 2007,[18] the Dalian PX protest in 2011,[19] the Shifang MoCu project protest in 2012,[20] and the Qidong protest on a paper mill's pollution discharge into the sea in 2012.[21] In August 2011, messages were widely spread through microblogs, blogs, Twitter, and other Internet forums that a paraxylene (PX) chemical factory (a joint venture between the city and a private company) built in Dalian city was at high risk to flood the town with the highly toxic chemical. Twelve thousand Dalian residents organized a peaceful public protest in Dalian's People's Square on August 14, demanding that the factory be immediately shut down and relocated and that the details about the investigation into the factory be made public. The Dalian government forbade the factory from opening.

According to statistics, the number of incidents caused by pollution increased at an annual rate of 29 percent between 2000 and 2006.[22] Those events raise challenges for preserving social stability, which is now another key target when evaluating local officials' performance. Mayors are thus becoming more concerned about local people's concerns for environmental quality and local quality of life. One way to reduce such mass protests is for mayors to preemptively address environmental challenges.

To measure urbanites' environmental concern and how it varies across different regions and cities, we have constructed two indexes to measure a province's residents' concern about environmental issues at a point in time. The first is the Google Insights index based on the Internet search intensity for the key words—*environmental pollution (huan jing wu ran)*.[23] Provinces with heavy industrial pollution (such as the northeast region, Chongqing, Shan'xi, and Yunan) have higher index values.

Based on our Google Insight indexes, we find that cities closer to Hong Kong and with more Internet users have higher public concern over pollution. The number of Internet users is a sign of the development of modern media, and cities close to Hong Kong are privileged with a relatively freer media environment; people there can access Hong Kong television and newspapers and also have some contacts with the people of Hong Kong. They have a better understanding of civil society and how the media works. In high-air-pollution cities, people use Google more as they search for information related to pollution. Cities with higher human capital have significantly higher public concern over pollution.

The Challenge of Corruption

Mr. Fang told us that local officials sometimes have a direct financial stake in factories or have personal relationships with the factories' owners. He thinks that such a relationship is not necessarily bad. "It can provide financing and connections for private firms, and can also help local officials to reconcile those firms' goals with their political goals such as economic growth, and sometimes also environmental protection," Mr. Fang told us. "In China's *guanxi* [social connection] society, local officials and firm managers need to make friends with each other. So long as the officials do not touch the corruption red line, it is okay." Mayor Han said that he has many good friends who are senior managers in the firms in his city; they sometimes dine together, and will discuss their mutual interests—how those firms can benefit both themselves and also the city. Mayor Han also asks them to help him to achieve his targets—the GDP number, tax revenue, and also environmental improvement. "If they are my friends, they will care more about my political career," he said. "They know that if I can be the mayor of this city for longer time, they can enjoy many conveniences I bring to them. This is not corruption, but just the reconciliation of our mutual long-term goals."

Both Mr. Fang and Mayor Han admitted that close relationships between local officials and firm managers may create conflicts of interest so that some local officials have less incentive to implement regulations that would improve the environment but would impose costs on the firms. In recent years, the Chinese media have uncovered cases in which local officials have put pressure on the courts, the press, or even hospitals to prevent the wrongdoings of factories from coming to light. Such officials also know that it is difficult for the central government to monitor and regulate their activities.

Corruption means that the rules do not apply and connected firms face less punishment. If this is true, these firms face less deterrence from environmental regulation and the profit maximizers will engage in more malfeasance. In some cases, local government acts as the polluting firms' "protective umbrella." The 2013 Milk River event in Kunming, Yunnan Province, in Southwestern China is a typical example. Mining firms along the Xiaojiang River discharged wastewater directly into the river and made it smell terrible. Due to the white color of the contaminated water, the local media dubbed it the Milk River. Farmers used this milky white water to irrigate their fields, and it destroyed many of their watermelons. But the director of local environmental protection administration simply denied the well-known fact that industry was responsible for the river pollution; this official was arrested at the end of 2013.

Professor Luo has some close relatives who are the senior managers in local firms in Guangdong Province. "Sometimes, the senior managers of those polluting firms are good friends with the officials in upper-level governments [such as the provincial government]," he told us. "Those officials then put pressure on local environmental protection agencies to overlook the firms' pollution." Local environmental agencies are responsible for pollution monitoring, but in this case corruption makes them turn a blind eye. "In many small cities in Guangdong Province, many people had protested against the polluting factories for a long time, but very few factories cared about this in the past," Professor Luo said. "The good thing is that, if the dirty firms are unfortunate enough to be included in the list of key polluting firms monitored by the central government, local officials cannot protect them anymore."[24]

Local government's punishments for polluting firms include fines, halting production, and shutting a firm down. Starting with the 1965 work on the economics of crime and punishment by Nobel laureate Gary Becker, economists have studied the key determinants of when a risk-neutral, profit-maximizing firm will comply with regulation. The simplest models of deterrence predict that a polluter will ignore the rules if the cost of complying with the rules is greater than the probability of being caught, multiplied by the dollar penalty imposed on violators. To give a specific example, suppose it would cost a firm $5,000 to install a piece of equipment to stop emissions. If the company does not install this piece of equipment, suppose that it faces a 10 percent chance of being caught not complying with emissions regulations and it will be fined $40,000 if it is caught. In this case, the firm will have a strong incentive to cheat because the $5,000 cost for complying is greater than .10 x 40000—that is, $4,000 for taking the gamble of not complying.

If the upper-level or central governments ask dirty firms to stop production or shut down, these firms will suffer a large loss. Some local governments comply with this order at first, but allow the firms to restart later when the media attention fades. Since in this case local officials share the risk with the polluting firms, it is likely that those firms may send more "gray money" to local politicians.

With the public's rising demand for political accountability and information transparency, as well as advances in modern media, such corruption today is more likely to be discovered. Mayor Shi told us, "I won't touch this corruption issue. I do not see significant benefits, but see enormous cost and danger—this will destroy my political career. I do not need money, and also do not want myself to become a bad star on Weibo." Mr. Fang explained that if an large environmental accident happens in a city, the city's mayor will face great pressure from his boss in upper-level government or even from the state. "In this case, even if you achieve very high GDP growth, you will have no chance to be promoted," he said. The combined pressure from the central government and the local people means that local officials are increasingly incentivized to address the major environmental challenges that their jurisdictions face.

Financing City Investment

Ambitious mayors seek to provide public goods and to build the infrastructure to create major cities, but this requires plenty of capital. Local governments need financial resources to build and maintain infrastructure such as public transit, sewers, and water treatment to allow a city to grow without exacerbating urban pollution problems. In Chinese cities, there is no property tax.[25] In the current tax-sharing system (*fen shui zhi*) established in 1994, the central government claims about half of all fiscal revenue,[26] but most financial obligations (infrastructure construction and public service provision) are still their responsibility.[27] A large share of local public expenditure is financed using transfers from the central government; in 2010 this share was 43.8 percent. Earmarked transfer payments from the central government are set for compensating local government expenditure on specific programs, such as education, health care, social security, and environmental protection. These transfer payments, and especially earmarks, play an important role in subsidizing specific local government expenditures. Antung A. Liu and Junjie Zhang find that such earmarks encourage the local government to invest in sewage treatment and thus reduces local water pollution.[28]

Another major revenue source for local governments to finance infrastructure comes from land sales.[29] Urban land is owned by the state, but in practice, the local (city) land bureau is responsible for the allocations of land through auction sales of leasehold rights and keeps the land-sale revenue.[30]

The central government has long prohibited municipalities from issuing bonds to finance government projects. To overcome such rules, cities set up so-called municipal investment companies to borrow hundreds of billions of yuan, mainly from state-run banks using land as collateral to finance infrastructure investments.[31] Western economists have estimated total local government debt in China at US$2 trillion to $3 trillion and rising. Large financial risks are believed to be associated with this infrastructure-financing option, which stems from municipal borrowing based on inflated land values offered as collateral to banks.[32] In 2013 the National Audit Office of China said that it would conduct a broad audit of debts incurred by local government agencies; this was the latest sign of

Beijing's concern that heavy borrowing on the part of local governments and their affiliates might pose a broader threat to the economy.[33]

Local governments' revenues tend to be heavily dependent on the sale to developers of long-term leases for government land, which is then used for building apartment and condominium complexes, factories, shopping malls, and other projects. Today, about one-third of the typical city's revenue is due to lump-sum land-sale revenue. Given the finite supply of land in cities, this public financing system is unlikely to persist. Recently the central government imposed strict regulations in China's urban housing market with the purpose of cooling down overheated property prices. Developers' interest in these leases tends to be cyclical, soaring when the real estate market is strong and crashing when real estate prices fall. With Beijing trying to make housing more affordable by limiting real estate speculation, developers have been more cautious about acquiring additional leases. This lowered the land-sale revenue that local governments received. "Our land-sale revenue dropped a lot in the last two years," Mayor Shi said, "but the expenditure does not decrease. Instead, it has expanded significantly because we want to improve public services such as education and health care. So I am really short of money now."

On the other hand, reliance on land-sale revenue in Chinese cities may also incentivize local governments to take local quality of life seriously because this will increase their land's value.[34] In several of our academic papers, we have documented how residential condominiums in Beijing and across China's major cities sell for a higher price premium if they come with access to green space, good air quality, and proximity to rapid public transit. Such a price premium for environmental amenities and a better quality of life provides an incentive for a forward-looking mayor to provide such public goods because such investments will raise the value of his remaining land inventory in the future.

While Chinese cities rely on land sales for revenue today, it is likely that cities will run out of land at some time point in the future. Farmers are growing much more assertive about their property rights to fringe land and, at the same time, many past urban manufacturing sites have already been converted and sold off in land sales. This will encourage such cities to introduce a property tax system. Those cities with high quality of life and high real estate prices could enjoy high revenues under this scheme.

To develop a more sustainable local public finance system, as pilot programs the Chinese central government allowed two cities to launch a property tax system in 2010 and further allowed four major Chinese cities to issue municipal bonds in 2011 (with the total bond amounts specified and monitored by the central government). If these pilot programs go well, such reforms are likely to expand to more cities. Mayor Shi understands that property tax is a better way to finance local public investment, but he does not think it can become law in the short run. "The discussion of property tax has lasted at least for ten years, but there are still lots of controversies," he said. "The key is who keeps this tax revenue—the city, the province, or the state? Of course I think the city should keep the revenue, because I spend money on those local public infrastructure projects." Other critics question how to effectively monitor the use of this huge tax revenue—if people think this money is likely to be misused, they will not support the increased tax burden.

Mayor Shi told us that in his city he uses the income from pollution charges to finance environmental infrastructure investment and operating costs. A special fund for pollution charges has been set up, and the charges collected from polluting firms are put into this fund. The polluting firms who plan to install emission reduction facilities or need subsidy for operating those facilities can apply for financing from this fund. Similar arrangements exist in other cities.

Local Government's Visible Hand in Influencing Land Supply and Firm Location

A distinctive feature in Chinese cities is that local governments have a visible hand in influencing firm location choices and urban expansion, with urban planning and land sales as the basic policy tools. Chinese local governments also have strong influence over determining the allocation of urban land supply through land sales. This affects the locational choices of firms and households.

Land taken from rural villages has fostered the growth of industrial zones in many suburbs. Local governments are keen to build large industrial zones at the city fringe with inexpensive land and favorable tax deduction policies to attract FDI and other firms that can produce high tax revenues. For example, in Zhejiang Province's "new technology zones," the government spent 100,000 yuan per mu (US$96,000 per acre) to provide basic infrastructure for the industrial land, but the average sale price of such industrial land to firms was only 86,000 yuan per mu (US$83,000 per acre). Half of the industrial land parcels were sold at a price less than 50 percent of the infrastructure construction cost. In some inland provinces that are eager to attract FDI and high-tax-revenue industries, some "new technology zones" sold their industrial land at a price of zero.[35] If those zones are hosts for high-tax but dirty industries, this will have a negative impact on local environmental quality. In addition, excess expansion of urban land at the city fringes increases the risk of urban sprawl and wasteful land use.

Outmigration

China's cities compete with each other, and such competition actually helps to protect the urban population's quality of life. As we argued in chapters 6 and 7, China's urbanites increasingly demand less risk in their lives. With the relaxation of the *hukou* internal passport limits, individuals will "vote with their feet" and move to those cities offering a better quality of life. Such credible migration restricts the ability of urban mayors in China to ignore the desires of their own people. Consider San Francisco; if that city's mayor sought to attract dirty industry in the name of accelerating local growth, air pollution would increase and quality of life would decline; many skilled people would leave, and home prices would fall. We predict that many cities in China will continue to be centers of manufacturing, but even more cities will choose to transition into modern service-sector cities whose industrial patterns mirror US cities like Chicago and San Francisco. As long as workers have full information about what quality of life is like in each city and have the ability to whatever city they choose, the menu of options protects workers against being "exploited" by any one mayor with a narrow focus on economic growth regardless of its quality-of-life impact. While international news coverage has focused on outmigration by Chinese elites to places such as Australia and western Canada, domestic Chinese migration will provide additional incentives for local officials to take the green agenda seriously. Those cities that ignore such demand for quality of life are likely to suffer a significant brain drain.

The leaders of China's major coastal cities appear to be well aware of the quality-of-life desires of their people. Shortly before the 2008 Olympic Games, Beijing required all polluting enterprises in the city center to stop operating or to move out. Construction sites also had to stop working, and the government implemented measures to limit the use of private vehicles according to their license plate numbers. During the preparation for the 2010 World Expo, the Shanghai government allocated about 3 percent of GDP to make investments in environmental protection. This money was spent to improve sewage discharge governance, to manage and relocate polluting enterprises, and to strengthen urban green construction. The city of Guangzhou also implemented emergency measures to reduce or suspend pollutant emissions for the 2010 Asian Olympic Games; these measures were effective in improving urban environmental quality in the short term—particularly air quality that is directly perceived by residents. But the question remains as to whether these measures will produce positive effects in the long term.[36]

CHAPTER TEN

Conclusion

Ms. Zhang is twenty-three years old. She comes from the city of Zheng-zhou, in Henan Province, and moved to Beijing in 2009 to study at Tsin-ghua University. Her parents still live in Zhengzhou, where they work as college professors. Although she has worried about Beijing's pollution, Ms. Zhang still decided to come to Beijing to benefit from its educational opportunities. Mr. Zhao is fifty-six years old. His hometown is in Jilin Province, in northeast China, and he grew up when China was quite poor. In his opinion, economic development is the most important objective to ensure that basic needs are met; environmental improvement is just the "icing on the cake."

This book has traced why Mr. Zhao was exposed to so much pollution during his lifetime while Ms. Zhang is likely to live in a much cleaner, high-quality-of-life city. While Mr. Zhao has worked in the same chemical plant for his whole life, Ms. Zhang is a member of a rising cohort of edu-cated young people starting their careers in urban China. This new gen-eration did not live during the Cultural Revolution. They have traveled to other nations, and are active readers of Internet news and blogs. Such experiences help create a vocal interest group seeking blue skies.

The rising demand for urban quality of life in China is fueled by the recognition that leisure, health, and happiness play a central role in de-termining an individual's quality of life. For decades, China's central gov-ernment policies focused relentlessly on income maximization. When Mr. Zhao was young, he chose to work in a chemical plant to earn money and to make a good living; he worked six days per week, and his spare time was very limited; he spent it watching TV. "In the old days, when we got married, we usually had the big four—that is to say, television, refrigerator, washing machine, and a voice tape recorder. Now, the big four has become smartphone, computer, car, and house," he comments.

The evidence we have reported in this book suggests that young Chinese people's conception of the "good life" is very similar to that of their US and western European peers. Rising educational attainment and per capita income in China will stimulate an increased demand for a cleaner urban environment. Labor economists studying human development over the life cycle have convincingly demonstrated that early life investments play a key role in later life success. Today, China is investing huge sums of money in building up the human capital of its young people.

Lower pollution levels increase the population's health and productivity, allow people to make greater use of the skills they have acquired, and allow them to enjoy their leisure time more. In this sense, pollution mitigation progress improves the standard of living by boosting productivity and nonmarket quality of life.[1] Human capital acquisition and good health complement each other: a child learns more in school when she is healthy. Anticipating a long life encourages forward-looking individuals to make greater long-run investments in human capital. Given that macroeconomists view human capital as the key input in urban economic growth, a cleaner environment thus directly contributes to China's long-run prosperity.

In an industrial economy, an economic boom leads to significant pollution as factories generate both products and emissions. In the modern service and high-tech economy, economic growth is stimulated by *lowering* pollution because the educated workforce is more productive in "blue sky" cities. In the 1980s Mr. Zhao and his friends were willing to work in the chemical plant because it paid enough money to feed their families. Today, a city with blue skies, good schools, and vibrant urban life attracts and retains talent.

At the same time that the demand for environmental progress is increasing in China's cities, there are economic forces at work that encourage dirty factories to move away from the major coastal cities. China's major cities will continue to deindustrialize as improvements in transportation networks, high land prices, high wages, and increasing environmental regulation encourage dirty factories to relocate to secondary cities. This industrial migration could cause increased pollution levels in the cities where heavy manufacturing grows, but this pessimistic view implicitly assumes that the new factories opening up will be just as dirty as the old factories that are closing. As demonstrated by the recent environmental history of the United States and western Europe, engineering progress and the enforcement of tightening environmental regulations raise the possibility that emissions per unit of production can decline enough to offset the scale effect of rising production levels.

Higher-Quality Urban Economic Growth

For decades, China's economic growth was anchored by heavy industry. Similar to England's industrial cities in the nineteenth century and Pittsburgh in the twentieth century, growing industrial cities experienced sharp increases in pollution.[2] A more balanced urban economy featuring a diversified industrial base that includes improved services and high-tech firms will enjoy higher quality growth. Macroeconomics students are taught to calculate a nation's net domestic product by deducting capital depreciation from GDP. In recent work, economists have attempted to incorporate this basic insight into their studies. As Mark L. Egan, Casey B. Mulligan, and Tomas J. Philipson write,

> Many national accounts of economic output and prosperity, such as gross domestic product (GDP) or net domestic product (NDP), offer an incomplete picture by ignoring, for example, the value of leisure, home production, and the value of health. Previous discussed shortcomings of such accounts have focused on how unobserved dimensions affect GDP levels but not their cyclicality, which affects the measurement of the business cycle. This paper proposes a new methodology to measure economic fluctuations that incorporates monetized changes in health of the population in the United States and globally during the past 50 years. In particular, we incorporate in GDP the dollar value of mortality, treating it as depreciation in human capital analogous to how net domestic product (NDP) treats depreciation of physical capital. Because mortality tends to be procyclical, we find that adjusting for mortality reduces the measured deviations of GDP from trend during the past 50 years by about 30% both in the United States and internationally.[3]

Their study is directly applicable to China today. Jinnan Wang has estimated that China's particulate levels caused 300,000 to 500,000 premature deaths each year in 2003–13. As the nation's urbanites grow wealthier and are willing to pay more to be exposed to less risk, this health capital cost caused by current levels of pollution will rise further.[4] This creates an imperative for the central government to address this important issue.

Recent macroresearch by Fergus Green and Nicholas Stern supports the optimistic hypothesis that China is increasingly prioritizing the quality of its growth, and this leads to a reduction in China's aggregate coal consumption. They point out that China has grown rapidly for more than three decades by following a strategy of high investment, strong export orientation, and energy-intensive manufacturing but that there have been large social costs to this growth, and the nation is now entering a maturation period in terms of skills, productivity, and rising wages:

China has now entered a new phase of economic development focused on better quality growth. From structural changes in the economy to explicit policies on efficiency, air pollution and clean energy, China's new development model is continuing to promote economic growth while driving down its GHG [greenhouse gas] emissions. . . .
Whereas coal consumption in China grew at around 10 percent per year in the first decade of this century, it fell in 2014 by 3 percent and into first quarter 2015. Use of natural gas in these sectors will increase rapidly over the next 5 to 10 years.[5]

Green and Stern predict that China's greenhouse gas emissions will peak between 2025 and 2030.

The Central Government's Willingness to Tackle the Pollution Challenge

Our optimism about increasing blue skies over China's cities hinges on central and local governments' increased willingness to tackle pollution problems. This shift in government priorities is due to a desire to please urbanites (who now represent a majority of the nation's population); the pursuit of a blue skies agenda helps incumbent government officials to consolidate their political power as they are seen as caring for the common people. Such quality-of-life efforts reduce the probability of future social unrest as the government establishes its legitimacy.

Recent macroeconomic research has examined how a nation's income distribution and long-run growth is determined by the rules that the nation's leaders introduce. In their book *Why Nations Fail*, Daron Acemoglu and James Robinson conclude that many nations fail to enjoy economic growth because their leaders introduce extractive institutions. Elites, when politically powerful, set up the "rules of the game" to capture resources and transfer them to themselves. These implicit taxes destroy the incentive for the majority of citizens in these nations to invest in human capital or other investments, and the net effect is economic stagnation.

The link between Acemoglu and Robinson's work and our study is that environmental improvements in cities represent inclusive public goods provision. Everyone in a city gains as air and water quality improve. Real estate prices will rise in such cities if only a few cities experience environmental progress and people can freely migrate from one city to another. This increase in real estate prices is less likely to happen if government policy causes environmental quality to improve in many cities. In this case, successful government provision of improved urban environmental quality helps to improve everyone's quality of life across China. Today in China there is great concern about rising income inequality. If the government's actions help to improve all people's quality of life, this helps to partially mitigate income inequality. The provision of public goods (clean air, food safety regulation, and clean water) is one strategy for broadly improving a population's quality of life and sharing the gains from economic development. A government that provides such public goods simultaneously increases the day-to-day quality of life and the productivity of its people, and such an inclusionary investment is likely to foster unity and political support.

While China's rise as an economic power has been based on industrial production and coal burning, in recent years the nation's major cities—such as Beijing and Shanghai—have enjoyed improvements in urban air pollution. Figure 10.1 shows the average particulate matter up to ten micrometers in size (PM_{10}) level in each of these cities between 2001 and 2013.

During these years, Beijing's PM_{10} levels have declined by 39 percent and Shanghai's by 20 percent. To achieve such ambient air-quality progress

PM$_{10}$ (ug/m^3)

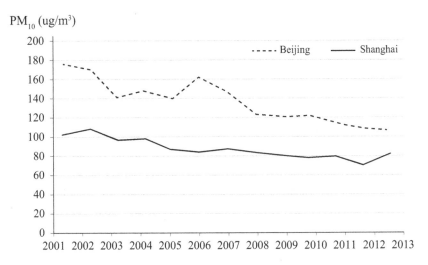

FIGURE 10.1 PM$_{10}$ in Beijing and Shanghai, 2001–13

during a time of growth in population, per capita income, and automobile use highlights how improvements in technique (emissions per dollar of local gross domestic product) can offset the pure scale associated with economic growth. Beijing's PM$_{10}$ levels remain much higher than cities of similar size, such as Los Angeles, but are now much lower than average levels in major cities in India.[6] As our examinations of manufacturing (chapter 2) and driving (chapter 5) highlight, there are a number of public policies that will further reduce emissions. Based on the political economy analysis we presented in chapters 8 and 9, we are optimistic that the leaders of these cities have incentives to implement many of these "green city" policies.

We recognize that not every Chinese city will experience such pollution mitigation progress. In cities such as Xingtai there is a tension between preserving the economic activity provided by heavy industry versus enjoying environmental gains. In such cities, government officials are wrestling with regulating large coal mines and shutting down or upgrading coal-fired power plants. To quote the *Wall Street Journal*, "Some of Xingtai's 7.6 million residents applaud the moves. But others—particularly those who work at Jizhong—are ambivalent. Pay is falling for some workers along with the company's profits, and many worry their jobs and the city's coal-based economy could be damaged. 'My monthly wages are down 50%,' said 25-year-old Shi Yang, who says he works fewer hours at a Jizhong coal-preparation plant." [7] This example highlights the trade-off that China now faces. Its wealthier cities along the coast are more likely to choose the "green path," than poorer cities. This Xingtai example highlights that the rise of China's green cities is not a "free lunch."

Environmental economists continue to try to quantify the costs of regulation and who bears those costs. When regulations are adopted that restrict how power is generated or force polluters to reduce their emissions, firms will earn less profit, consumers will pay higher prices, or workers will be less likely to be employed in that industry. The distribution of such costs has important distributional implications for who pays for China's blue skies.

Our data analyses allow us to make a more precise prediction of how many of China's cities fall into the Beijing and Shanghai rapid-progress group versus the slow-progress Xingtai group. Over the last twenty years, academic environmental economists have debated a hypothesis called the environmental Kuznets curve (EKC, see Figure 1.1).[8] This is not esoteric academic debate. The EKC hypothesis posits that economic growth is both a *foe* and a *friend* of the environment. In particular, the EKC is a statistical claim that as poor cities and nations grow wealthier they will suffer environmental degradation, but beyond a per capita income *turning point* that economic growth and pollution progress will be positively correlated and hence economic growth actually stimulates environmental progress.

An entire subfield of environmental economics has emerged that focuses on this inverse U association between per capita income and pollution. In past research, we have estimated the EKC across China's cities.[9] Here we focus on eighty-three Chinese cities for which we can access ambient air pollution (PM_{10}) data. This set of cities includes medium and large cities as well as some small ones. Using standard statistical regression methods, we have studied PM_{10} dynamics for the cities over the years 2003–12 as a function of the city's per capita income dynamics, population size, human capital, industrial composition, and climate. Appendix 1 reports our EKC equation specification and regression results. The results of the EKC regression illustrate that cities with a larger population and a large manufacturing share are more polluted, and cities experiencing less rainfall are more polluted. Controlling for these factors, we focus on the role of a city's per capita income. Based on our regression estimates, we find that wealthier cities are enjoying PM_{10} mitigation progress and that

TABLE 10.1 Predicted air pollution dynamics for Chinese cities

Year	Number of cities that pass the EKC turning point	Total urban population size in those cities (in millions)	Average annual decline rate in PM$_{10}$	The top five cities that enjoy the largest PM$_{10}$ decline
2010	33 (out of 85)	134.96	—	—
2020	56 (out of 85)	367.61	–3.4%	Suzhou, Guangzhou, Ningbo, Dalian, Xiamen
2030	78 (out of 85)	682.55	–4.9%	Zhuhai, Guangzhou, Xiamen, Suzhou, Ningbo

the key "turning point" occurs at 84,000 yuan (about US$13,000) in terms of GDP per capita. Intuitively, those cities whose per capita income is greater than 84,000 yuan are predicted to enjoy PM$_{10}$ mitigation progress as they grow wealthier.

By 2012 there were already thirty-three cities (out of eighty-five) that had passed this turning point, and about 140 million urbanites lived in them. Our model also allows us to generate some predictions about future urban pollution levels based on different assumptions about China's long-term annual per capita GDP growth rate. In one scenario, we assume that all cities will experience annual per capita income growth of 5.2 percent over the next twenty years. Table 10.1 reports that according to this scenario, by the year 2020, twenty-three more cities will pass the turning point, and another twenty-two cities by 2030. If we count the urban population in those cities, 231 and 315 million Chinese urbanites will live in those cities with significant air quality progress in 2020 and 2030, respectively. For those "greening" cities, the average annual PM$_{10}$ decline rates are 3.4 percent and 5 percent during these two periods. This is a large effect. In the table we also list the top five cities that will enjoy the largest PM$_{10}$ declines. They are Suzhou, Guangzhou, Ningbo, Dalian, and Xiamen for 2012–20; and Zhuhai, Guangzhou, Xiamen, Suzhou, and Ningbo for 2020–30. We also analyze a second scenario in which all cities will grow in the future based on their per capita GDP growth rate between 2005 and 2014. This scenario yields an even more optimistic result: sevety-nine of the eighty-five cities will pass the turning point by 2020.

Figure 10.2 displays the spatial distribution of the Chinese cities for which we predict near-term improvement in ambient air quality.

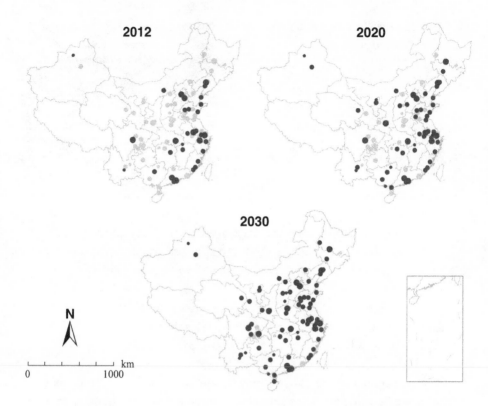

FIGURE 10.2 Cities that pass the environmental Kuznets curve turning point
Note: Dark dots are those cities that have passed (in 2012) and will pass (in 2020 and 2030) the turning points; gray dots are cities that lag behind; larger dots represent larger cities (in terms of population).

This empirical evidence boosts our confidence in our prediction that hundreds of millions of Chinese urbanites will enjoy an improvement in their ambient air quality. This will offer both aesthetic quality-of-life benefits and direct productivity benefits. We recognize that these predictions merit future verification.

Testing Our Core Claim

This book's core claim is that many major Chinese cities will experience environmental improvement in the medium term. We would reject our hypothesis if over the next two decades we could observe the following pieces of evidence. The first piece of evidence would be based on objective trends in ambient air and water pollution: if we observe pollution to be rising in cities where we predict that it should be dropping. Beijing's ten-year trend in ambient particulate levels indicates significant air quality improvement during a time of population and per capita income growth. Such a pattern could only be observed if pollution per unit of economic activity (a technique effect) is improving faster than the scale of economic activity is rising. The second piece of evidence would be a reversal of course on the part of the Chinese central government and a return to a official promotion criteria based solely on local GDP growth; such an incentive program would encourage mayors to pursue growth over any side goals of sustainability. The third piece of evidence would be better-educated and wealthier cities in China seeking to retain dirty industrial production rather than transitioning to a local economy based on services. The fourth piece of evidence would be a continued effort on the part of the Communist Party to keep energy prices low and to continue to rely on coal-fired power. The fifth piece of evidence would be urban residents themselves no longer voicing demands for environmental protection and access to trusted information. The sixth piece of evidence would be continued investments on the part of China's firms and government in coal-fired power plants and other durable, long-lived investments that offer economic resources but degrade the environment. The seventh piece of evidence would be media suppression of Chinese environmental issues. Today, one cannot access the *New York Times* website in China. If the environmental media and social media chatting on urban quality-of-life issues is suppressed and unable to play the role, industrial malfeasance is more likely to take place and externalities will be less likely to be mitigated.

How Much Does China's Growth Exacerbate Climate Change Risk?

This book's main focus has been on day-to-day quality of life for Chinese urbanites. We have studied how their individual choices and how the choices of China's urban firms aggregate to determine local pollution levels in China. In a city featuring millions of people and thousands of firms, no one individual's actions has a measurable impact on the environment.

If coauthor Siqi Zheng bikes to work rather than drives, this reduction in emissions has a miniscule impact on Beijing's air pollution. Urbanites such as Siqi are simultaneously producers of urban pollution and consumers of this public ill. China's urbanites have strong incentives to know the daily urban air pollution levels and to take protective actions (such as staying inside or wearing an air mask).

At the same time that urbanites both supply pollution and try their best to avoid local pollution, another unintended consequence of daily urban life is to produce greenhouse gas emissions and hence in aggregate to exacerbate the risk of global climate change. China is a very large and growing economy, and the nation's urban growth has implications for global greenhouse gas production.

Here we present a simple forecasting exercise to predict how the electricity generation sector and rising private vehicle use in China will contribute to global carbon dioxide (CO_2) emissions by the year 2040, and we make predictions for China and the United States. In conducting this speculative exercise, we only consider the scale effect (the total size of electricity consumption) and the composition effect (the composition of energy sources to generate that amount of electricity), but we are unable to forecast the changes in emission factors of those energy sources (technique effect) in the long term. We therefore base our predictions using emissions factors from the present day. China's National Energy Administration released a report on the medium- to long-term electricity consumption demand in China. We borrow its forecast for China's electricity consumption in 2040 and the composition change of fuel sources to generate that amount of electricity. Similarly, the US Energy Information Administration forecasts US electricity consumption number in 2040 and the composition of the fuel sources.

Table 10.2 shows that the electricity demand will roughly triple from 2010 to 2040 in China due to income growth and urbanization. Thanks to the major switch from coal to natural gas and other clean energy sources (the share of coal will decrease from 79 percent to 57 percent), total CO_2 emissions in 2040 will just be 2.2 times those of 2010 based on current emission factors. The table highlights that in the case of CO_2 emissions, the scale effect of rising electricity demand in China will overwhelm the environmentally beneficial composition effect of China moving away from

TABLE 10.2 Forecasting CO$_2$ emissions from the electricity generation sector in the year 2040

	China: 2010	China: 2040	US: 2010	US: 2040
Population (in millions)	1360	1435	312	383
Electricity consumption per capita (KWH)	3094	8869	13195	14615
Total electricity consumption (in billions of KWH)	4207	12731	4120	5600
Electricity generation by sources				
Coal	79.10%	57.00%	45.80%	32.00%
Natural gas	1.36%	4.10%	23.38%	35.00%
Oil	0.44%	0.00%	1.10%	1.00%
Other clean energy sources	19.10%	38.90%	29.72%	32.00%
CO$_2$ emission (million tons)	3366.50	7477.75	2329.45	2664.27

Sources: Population data is collected from the United Nations Department of Economic and Social Affairs, Population Division, *World Urbanization Prospects 2014*, http://esa.un.org/unpd /wup/highlights/wup2014-highlights.pdf; the source shares for 2010 are collected from World Bank indicators, http://data.worldbank.org/indicator); others are collected from China's National Energy Administration as well as the U.S. Energy Information Administration, *Annual Energy Outlook 2015*, http://www.eia.gov/forecasts/aeo/er/early_elecgen.cfm.

dirty coal toward cleaner natural gas and renewable sources of power. Obviously, if China generates more of its power from renewables in the year 2040, its total CO$_2$ emissions will decline. We will return to this point in the next section.

Greenhouse Gas Emissions from Transportation

We repeat the forecasting exercise for cars, to see how many cars will be driven on the roads in Chinese cities in 2040, and to predict the total CO$_2$ emissions from those cars. We base these predictions on a simple algebra equation as expressed by $(N \times P \times M \times E)$, where N is urban population, P is the probability of an urbanite owning a vehicle, M is miles driven per car, and E is emissions per mile. E is calculated using miles per gallon and the CO$_2$ emission factor per gallon of gasoline. Table 10.3 shows our forecast's results.

TABLE 10.3 CO$_2$ emission from private cars in China and the United States in 2010 and 2040

Nation	China				United States	
Year	2010	2040 (Scenario 1)	2040 (Scenario 2)	2040 (Scenario 3)	2010	2040
N (million)[1]	669.4	1044.4	1044.4	1044.4	252.2	329.0
P (% vehicle ownership)[2]	5%	30%	35%	40%	75%	85%
M (annual miles driven)[3]	9320.6	11000.0	11000.0	11000.0	12359.0	12800.0
MPG (miles per gallon of gasoline)[4]	24.0	46.0	46.0	46.0	32.7	55.6
E (kilograms CO$_2$ per mile)	0.593	0.310	0.310	0.310	0.435	0.256
Transportation CO$_2$ emissions per year (million tons)	185.1	1066.7	1244.5	1422.3	1017.8	916.7

Note: China's MPG data is assumed to be equal to the efficiency of the world's average.

Sources: [1] United Nations, Department of Economic and Social Affairs, Population Division *World Urbanization Prospects 2014*, http://esa.un.org/unpd/wup/highlights/wup2014-highlights.pdf.
[2] National Bureau of Statistics of China, *China Energy Statistical Yearbook 2010* (Beijing: China Statistics Press, 2010); US Department of Transportation, Federal Highway Administration, *2009 National Household Travel Survey*, http://nhts.ornl.gov; Kenneth B. Medlock III, Ronald Soligo, and James D. Coan, *Vehicle Stocks in China: Consequences for Oil Demand*, http://bakerinstitute.org /files/530/.
[3] US Energy Information Administration, *Annual Energy Outlook 2014 with Projections to 2040*, http://www.eia.gov/forecasts/aeo/pdf/0383(2014).pdf; Ira Sohn, "Fueling China's Growing Automobile Density Rate: Some Back-of-the-Envelope Calculations for the 2030–2050 Interval," *Indian Journal of Economics and Business* 9, no. 4 (2010): 861–74.
[4] US Energy Information Administration, *Annual Energy Outlook 2014 with Projections to 2040*, http://www.eia.gov/forecasts/aeo/pdf/0383(2014).pdf; ExxonMobil, *The Outlook for Energy: A View to 2040*, http://corporate.exxonmobil.com/en/energy/energy-outlook.

There is no reliable forecast regarding the car ownership rate (cars per hundred people) in Chinese cities for the year 2040. Some experts say it will be around 30–40 percent, depending on whether more cities will impose the strict demand-side regulation policies such as Beijing's driving and car purchase restrictions. We therefore list three scenarios in table 10.3 with assumed car ownership rates of 30 percent, 35 percent, and 40 percent. Rapid urban expansion also means that each driver will drive more (11,000 miles per year) than he or she does now. The scale effect of more drivers traveling more miles can only be offset if future vehicles' miles-per-gallon rates rise sharply. We assume that the average vehicle achieves forty-six miles per gallon in the year 2040 versus twenty-four miles in 2010.

Combining all the above assumptions together yields our prediction that China's annual transportation CO_2 emissions will rise from 185 million tons to 1,067 million tons, 1,245 million tons, and 1,422 million tons under the three scenarios between 2010 and 2040. In the United States, both the urban population and car ownership growths are quite moderate, and these two effects are offset by the technique improvement. This algebra highlights that China's vehicle growth has major consequences for exacerbating climate change. No one driver has an incentive to recognize how his emissions contribute to the problem because each individual only contributes a tiny fraction of the overall emissions. This algebra provides a salient example of the natural resource "consumption factor" associated with economic growth. Environmentalists such as Jared Diamond have warned that the convergence in world per capita income between poor nations and wealthy countries could have drastic environmental implications.[10]

Both the electricity demand scenarios presented in table 10.2 and the aggregate gasoline demand scenarios presented in table 10.3 suggest that China's CO_2 emissions will continue to rise sharply over the next twenty-five years. But it must be pointed out that these linear extrapolation exercises abstract away from incorporating microeconomic incentives for technological innovation. Innovations that lower the cost of generating renewable power and innovations that accelerate the deployment of electric vehicles could sharply reduce the carbon intensity of China's growing economy.

Can Greenhouse Gas Emissions Decline in Urban China?

The predictions presented above based on an extrapolation exercise suggest that China's greenhouse gas emissions will rise sharply in the upcoming decades. Such exercises implicitly assume a business-as-usual policy approach. If China's government were to commit to a carbon mitigation strategy such as introducing a carbon tax, the carbon emissions factor from power generation and private vehicle transportation would fall sharply. Researchers such as John A. Mathews and Hao Tan have argued that China's carbon intensity of its electricity production is starting to quickly decline as it has increased its share of power generated by renewables.[11]

As we noted in chapter 2 and table 10.1, China continues to rely on coal for power generation. Environmental health researchers have often pointed out the benefits associated with reduced coal reliance. When coal is burned, this activity has consequences for the global externality of climate change and also elevates local particulate levels. A cobenefit of reducing coal reliance would be a reduction in local particulate levels in China. A recent International Monetary Fund working paper estimates that such cobenefits are large—equal to sixty-three dollars per ton of CO_2 released.[12] Given that China's per capita income is rising, and that wealthier people seek to avoid risk, this cobenefit of reducing CO_2 production will rise over time.[13] This means that China has an increased private incentive as a nation to consider ways to reduce its greenhouse gas production.

Given the scale of Chinese electricity consumption and driving, the only path for China to take to significantly reduce its greenhouse gas emissions would be to sharply increase the share of its electricity generated by renewables and natural gas and for its new vehicles to be mostly electric vehicles rather than those powered by gasoline. A transition away from heavy manufacturing to services would also lead to lower overall electricity demand in China.

Technological change at the global level will play a key role in decoupling the impact of economic growth on greenhouse gas production. Many academic economists are actively researching directed technological change.[14] This research seeks to understand which problems entrepreneurs and venture capitalists should focus their efforts on trying to solve. As highlighted by the 2015 United Nations Climate Change Conference in Paris, all of the nations in the world are converging on introducing hard and soft caps on their greenhouse gas emissions. While many details need to be worked out, these international negotiations signal to entrepreneurs that there is a large nascent market demanding low-carbon sources of energy and consumer durables that are energy efficient.

While there is much we do not know about the challenge of climate change, China does face some serious risks posed both to agriculture production and to cities in areas that will experience sea level rise and in areas that could experience extreme summer heat. Given that the United States and China are responsible for roughly 40 percent of the world's greenhouse gas emissions, these two nations working together could take actions that would have the net effect of slowing down the growth of global emissions. In his recent American Economic Association Presidential Address, William Nordhaus of Yale University discussed the incentives that different nations have out of self-interest to participate in such an agreement. He argues that "climate clubs" may emerge in which subsets of major nations agree to commit to new rules that help reduce global emissions and thus act as an insurance policy reducing the ugliest scenarios associated with climate change. Nordhaus writes,

> The club is an agreement by participating countries to undertake harmonized emissions reductions. The agreement envisioned here centers on an "international target carbon price" that is the focal provision of an international agreement. For example, countries might agree that each country will implement policies that produce a minimum domestic carbon price of $25 per ton of carbon dioxide (CO_2). Countries could meet the international target price requirement using whatever mechanism they choose—carbon tax, cap-and-trade, or a hybrid. A key part of the club mechanism (and the major difference from all current proposals) is that nonparticipants are penalized. The penalty analyzed here is uniform percentage tariffs on the imports of nonparticipants into the club region. An important aspect of the club is that it creates a strategic situation in which countries acting in their self-interest will choose to enter the club and undertake high levels of emissions reductions because of the structure of the incentives.[15]

While this is just one suggestion for how the United States and China can work together to mitigate climate change, it highlights the key role that environmental economics plays in informing international relations. Both China's and the United States' urban quality of life would be protected or improved if such a deal could be formalized.

On November 12, 2014, China and the United States unveiled new pledges on greenhouse gas emissions as the leaders of the two countries met for talks during the 2014 Asia-Pacific Economic Cooperation Summit in Beijing. The deal commits China to reducing greenhouse gas emissions after a peak in 2030 (and ideally sooner). China also aims to have nonfossil fuels make up 20 percent of its primary energy consumption by 2030.[16] It is the first time China, the world's largest emitter by far in absolute terms (roughly 28 percent of the world's CO_2 emissions in 2014), has agreed to set a ceiling—albeit an undefined one—on overall emissions. This target was "notable," a White House statement reads. "It will require China to deploy an additional 800–1,000 gigawatts of nuclear, wind, solar and other zero-emission generation capacity by 2030—more than all the coal-fired power plants that exist in China today and close to total current electricity generation capacity in the United States. Mr. Obama called the agreement 'historic' and said the US would work with China to 'slow, peak and then reverse the course of China's carbon emissions.'"[17]

In September 2014 it was announced that China will experiment with a carbon cap-and-trade market starting in 2016.[18] The introduction of a credible cap-and-trade carbon pollution permit market would accelerate the rate of technological progress and diffusion of electric vehicles and renewable power generation. According to the current implementation plan, the cap will include power plants and heavy industry. The central government will be able to monitor how agile different polluters are in adapting to these new rules. While many environmentalists claim that the costs of compliance are low and many US Republicans claim that the costs of carbon abatement are enormous, such a field experiment by the Chinese will provide accurate real-time estimates, and this new information will allow the central government to then make a decision on how to adjust the new regulation's details to balance achieving environmental gains against sacrificing economic growth. This iterative process will help China reduce its greenhouse gas emissions.

The extra cost for generating power using renewables rather than coal is not a law of physics. The cost of clean, renewable power will decline over time as innovation takes place around the world in the building of more efficient solar panels and wind turbines. At his January 2015 inauguration California governor Jerry Brown announced the goal of California achieving 50 percent of its power from renewables by the year 2050.[19] This major state committing itself to a "moon shot" goal represents a commitment device: entrepreneurs such as Elon Musk observe this commitment and now can more confidently make plans to invest in risky research and development knowing that California's electric utilities will buy green power if the entrepreneurs can produce it at relatively low cost. This big push for achieving green innovation will only be accelerated if a huge market such as China also commits to renewable portfolio goals.

China's urban growth actually helps to stimulate technological progress because a larger market of potential demanders creates an incentive for entrepreneurs to invest in costly and risky research and development to attempt a breakthrough. If Chinese middle-class households are willing to pay for energy efficient air conditioners, this creates a huge market payoff for the global firm that can deliver such a project. The awareness of this opportunity creates a strong incentive for global firms to take risks in trying to design such products.

At the same time, China's ongoing efforts to identify new export markets has led its layers of government to make strategic investments in free land and low interest rates for producers of wind turbines and solar panels. As these exports improve in quality and drop in price they play a key role in global supply chains, helping to make a green economy an affordable reality. For example, Walmart has committed to increasing its reliance on renewable power.[20] If the price of solar panels produced in China declines (due to global competition and learning by doing), and if their quality increases, companies such as Walmart are likely to accelerate their investments in such technology that shrinks the firm's carbon footprint.

How much this cost will decline remains an open question, but China's demand for such green products and its active pursuit of being a major exporter of this nascent technology increases the likelihood of a major low-carbon breakthrough that can be mass-produced and used all over the world.

If nations such as China and the United States see that the cost of reducing carbon emissions is falling over time, this reduces the cost of voting in favor of an international carbon tax. While China's citizens do not directly vote, leaders are likely to introduce policies that offer benefits if the cost of the policy falls. In the case of the United States, voting research both at the congressional level and in California indicates that voters are more likely to oppose carbon mitigation regulation if they face a higher out-of-pocket expense from such regulation. These facts are based on an analysis of congressional voting on the American Clean Energy and Security Act in 2009 and on voting data in California on Proposition 23 in 2010, a proposition that would have suspended California's low-carbon Global Warming Solutions Act of 2006 dictated a carbon cap-and-trade market for the state.[21] This research finds that suburban residents, who produce more greenhouse gas emissions and are aware of this fact, tend to vote with their pocketbooks and oppose carbon pricing. If new technologies decoupled such lifestyle from carbon production, suburban middle-class support for low-carbon policies would rise.[22]

Final Thoughts: Life Satisfaction Dynamics in Urban China

Richard Easterlin and his colleagues have researched life satisfaction in different nations around the world. Based on this team's recent survey evidence in China, they conclude, "Despite its unprecedented growth in output per capita in the last two decades . . . there is no evidence of an increase in life satisfaction of the magnitude that might have been expected to result from the fourfold improvement in the level of per capita consumption that has occurred."[23]

As we have discussed throughout this book, modern Chinese urbanites face great stress from a combination of sources including workplace expectations, home prices, local quality of life, and concerns about their children's future. Easterlin and colleagues' surveys are likely to reflect these concerns. In a nation experiencing fast per capita income growth, the standard of living has not risen as quickly in part because of lingering pollution challenges.

Given China's unique history and political institutions, we cannot extrapolate from our findings about the nation to make informed predictions about the consequences of urban growth for other developing nations such as India, Indonesia, and Vietnam. We do believe that the aspirations of the urban, educated middle and upper classes in each of these nations are likely to be quite similar. Each of these nations will wrestle with the trade-off of relying on inexpensive but dirty fuels such as coal versus moving up the energy ladder. Each of these nations will face choices over the industrial comparative advantage versus the environmental implications of a city's industrial composition. Similar to China's case, as urbanites grow in number and grow wealthier, there will be an increased demand for private space and private vehicle use. The impact on local and global environmental challenges will depend on whether these nations adopt cutting-edge technologies and how much their residents are willing to pay for improvements in environmental quality. As we learn from the China case, government incentives to address local and global environmental concerns play a key role in mediating the consequences of urban growth on the environment. The incentives for leaders to step up needs to be studied on a nation-by-nation basis.

We use equation A1 to estimate the relationship between a Chinese city's ambient particulate matter (PM_{10}) level and the city's characteristics, including its per capita income. Following the academic literature studying the environmental Kuznets curve, we include a polynomial of urban per capita income. This allows us to test for whether there is a "turning point" such that when a city's per capita income exceeds this turning-point level there is a negative association between economic growth and pollution. We find that a third-order polynomial expansion of urban gross domestic product (GDP) per capita produces the highest adjusted R^2. PM represents annual mean PM_{10} concentration. City population and education level in 2000, the manufacturing share, temperature index, and rainfall are included as controls ($\mathbf{X}$).

$$Ln\left(PM_{it}\right) = \eta_0 + \sum_{j=0}^{J}\eta_{1j} \times [Ln\left(GDP\ per\ capita_{it}\right)]^j + \mathbf{\eta_2} \times \mathbf{X}_{it} + \lambda_t + \xi_{it} \qquad A1$$

Table A.1 reports the regression result of estimating equation A1.

Based on the regression result in table A.1, we find that wealthier cities enjoy PM_{10} mitigation progress and that the key turning point occurs at 64,000 yuan (about US$10,000) in terms of GDP per capita.

TABLE A.1 City-level environmental Kuznets curve regression

Dependent variable (in PM$_{10}$ concentration)	EKC regression
ln GDP per capita	−2.230*
	(0.514)
ln GDP per capita, squared	0.542*
	(0.149)
ln GDP per capita, cubic	−0.044*
	(0.014)
ln population, 2000	0.169*
	(0.014)
ln education level, 2000	−0.131*
	(0.122)
Manufacturing share in GDP	0.0062*
ln GDP per capita	(0.00088)
Constant	−0.986
	(0.657)
Other control variables	temperature discomfort index, rainfall, latitude, longitude
Observations	825
R^2	0.462

Note: Standard errors are reported in parentheses; *significant at the 1% level.

APPENDIX 2

The regression presented below estimates for eighty-three Chinese cities over the years 2005 to 2010. The dependent variable is a city's particulate matter (PM_{10}) concentration (in logarithm) in a given year, and the independent variables are the city's population (in logarithm), its industrial composition measured by the share of manufacturing jobs, and a time trend (representing technical change and public policy changes). The results are shown in table A.2.

TABLE A.2

Dependent variable *ln PM_{10}*	*(1)* *All cities*	*(2)* *High-income cities*
ln population	0.125*	0.212*
	(9.19)	(11.76)
Manufacturing share	0.249*	0.624*
	(2.62)	(4.78)
Year trend	−0.0282*	−0.0181*
	(−8.92)	(−4.33)
Constant	−3.011*	−3.839*
	(−31.08)	(−27.98)
Observations	938	499
R^2	0.146	0.236

Notes: *t* statistics in parentheses; *significant at the 1% level.

NOTES

Chapter 1: Introduction

1. Michael Greenstone, "Pollution around the World: A Matter of Choices," *New York Times*, December 30, 2014, http://www.nytimes.com/2014/12/31 /upshot/pollution-around-the-world-a-matter-of-choices.html?rref=upshot.
2. "Air Quality Suffers Due to Smog," *China Daily*, January 14, 2013, http:// www.chinadaily.com.cn/china/2013-01/14/content_16115953.htm.
3. Alex Wang, Orville Schell, Elizabeth Economy, Michael Zhao, James Fallows, and Dorinda Elliott, "Airpocalypse Now: China's Tipping Point?," *China File*, February 6, 2013, http://www.chinafile.com/airpocalypse-now-chinas -tipping-point. Particles 2.5 microns or less in diameter ($PM_{2.5}$) are referred to as "fine" particles and are believed to pose greater health risks than larger particles because they can become embedded deep in people's lungs.
4. World Bank, *World Development Indicators 2007* (Washington, DC: World Bank, 2007).
5. World Bank, *Cost of Pollution in China* (Washington, DC: World Bank, 2007).
6. Edgilis, *Outdoor Air Pollution in Asian Cities: Challenges and Strategies—Hong Kong Case Study* (Singapore: Edgilis, 2009).
7. "Environment May Be Issue at Two Sessions," *China Daily*, February 26, 2013, http://usa.chinadaily.com.cn/china/2013-02/26/content_16255385 .htm. Jennifer Duggan, "China's Environmental Problems Are Grim, Admits Ministry Report," *Guardian*, June 7, 2013, http://www.theguardian.com /environment/chinas-choice/2013/jun/07/chinas-environmental-problems -grim-ministry-report.
8. World Bank, "CO2 Emissions (Metric Tons per Capita)," http://data.world bank.org/indicator/EN.ATM.CO2E.PC/countries.
9. Matthew E. Kahn and Joel Schwartz, "Urban Air Pollution Progress despite Sprawl: The "Greening" of the Vehicle Fleet," *Journal of Urban Economics* 63, no. 3 (2008): 775–87.
10. In December 2013, 104 cities in twenty provinces in and near the Beijing-Tianjin-Hebei region and the Yangtze River delta region were overcome with a heavy haze that reduced visibility to less than ten meters in some places. See: "Cities Hit Hard by Smog," *China Daily*, http://usa.chinadaily .com.cn/china/2013-12/09/content_17160658.htm.

11. "China Readies Itself for CO2 Emissions Cap," *China Daily*, August 29, 2014, http://europe.chinadaily.com.cn/epaper/2014-08/29/content_18510113.htm.

12. Despite the film's demonstrating the failure of China's regulations on pollution, the Chinese government did not at first censor it. But within a week, the Communist Party's Publicity Department confidentially ordered the film to be removed. See Wikipedia, "*Under the Dome* (Film)," https://en.wikipedia.org/wiki/Under_the_Dome_(film).

13. Our definition of the middle class and upper middle class is based on Dominic Barton, Yougang Chen, and Amy Jin, "Mapping China's Middle Class," *McKinsey Quarterly*, June 2013, http://www.mckinsey.com/Insights/Consumer_And_Retail/Mapping_Chinas_middle_class?cid=china-eml-alt-mip-mck-oth-1306.

14. BP, *BP Statistical Review of World Energy June 2014*, http://www.bp.com/con tent/dam/bp/pdf/Energy-economics/statistical-review-2014/BP-statistical-review-of-world-energy-2014-full-report.pdf.

15. UK Foreign and Commonwealth Office, *China: Central Government Vision on Hukou Reform*, https://www.gov.uk/government/publications/hukou-reform-central-government-sets-out-a-vision-august-2014/hukou-reform-central-government-sets-out-a-vision-august-2014.

16. With origins in the central planning era, the five-year plans are still the most important policy instruments in China. Since each plan sets economic and social development goals for the next five years, major changes in policy orientation are usually reflected therein.

17. Siqi Zheng, Matthew Kahn, Weizeng Sun, and Danglun Luo, "Incentives for China's Urban Mayors to Mitigate Pollution Externalities: The Role of the Central Government and Public Environmentalism," *Regional Science and Urban Economics* 47 (2014): 61–71.

18. Genia Kostka, *Barriers to the Implementation of Environmental Policies at the Local Level in China*, Policy Research Working Paper WPS 7016 (Washington, DC: World Bank, 2014).

19. Gene M. Grossman and Alan B. Krueger, *Economic Growth and the Environment*, NBER Working Paper 4634 (Cambridge, MA: National Bureau of Economic Research, 1994).

Chapter 2: Made in China

1. Ryan Rutkowski, "Rebalancing and Rising Electricity Prices in China," China Economic Watch, Peterson Institute for International Economics, February 18, 2014, http://blogs.piie.com/china/?p=3760.

2. The Information Technology Authority of the Beijing municipal government signed a strategic partnership agreement with its counterpart in Hebei Province in August 2015 to build a cloud-computing industrial park in Zhangbei County. The county government said it would offer preferential policies such as lower taxes and electricity prices to attract companies to invest in the park. See Meng Jing, "Beijing, Hebei Team Up for Cloud Computing Park," *China Daily*, August 22, 2014, http://www.chinadaily.com.cn/beijing/2014-08/22/content_18468007.htm.

3. National Bureau of Statistics of China, "Consumption of Energy by Sector (2011)," http://www.stats.gov.cn/tjsj/ndsj/2013/indexeh.htm. US Energy Information Administration, "Consumption and Efficiency," http://www.eia.gov/consumption/.

4. Nicholas Z. Muller, Robert Mendelsohn, and William Nordhaus, "Environmental Accounting for Pollution in the United States Economy," *American Economic Review* 101, no. 5 (2011): 1649–75.

5. Tang Deliang, Tin-yu Li, Jason J. Liu, Zhi-jun Zhou, Tao Yuan, Yu-hui Chen, Virginia A. Rauh, Jiang Xie, and Frederica Perera, "Research: Children's Health," *Environmental Health Perspective* 116 (2008): 674–79.

6. MIT News Office, "Innovative Study Estimates Extent to Which Air Pollution in China Shortens Human Lives," http://web.mit.edu/newsoffice/2013/study-estimates-extent-to-which-air-pollution-in-china-shortens-lives-0708.html.

7. William Wan and Abigail Hauslohner, "China, Russia Sign $400 Billion Gas Deal," *Washington Post*, May 5, 2014, http://www.washingtonpost.com/world/europe/china-russia-sign-400-billion-gas-deal/2014/05/21/364e9e74-e0de-11e3-8dcc-d6b7fede081a_story.html.

8. Sarah Lain, "The Significance of the China-Russia Gas Deal," *Diplomat*, May 24, 2014, http://thediplomat.com/2014/05/the-significance-of-the-china-russia-gas-deal/.

9. Natural gas has a sulfur dioxide (SO_2) emissions factor of .057 (pounds per megawatt-hour of power generated) and an emissions factor of .505 for mono-nitrogen oxides (NO_x). In comparison, coal's respective emissions factors are 7.718 for SO_2 and 2.771 for NO_x. Electricity generated using natural gas has a 50 percent lower carbon dioxide emissions factor than electricity generated using coal. US Environment Protection Agency, "Clean Energy," http://www.epa.gov/cleanenergy/energy-resources/egrid/index.html.

10. Cesur Resul, Erdal Tekin, and Aydogan Ulker, *Air Pollution and Infant Mortality: Evidence from the Expansion of Natural Gas Infrastructure*, NBER Working Paper 18736 (Cambridge, MA: National Bureau of Economic Research, 2013).

11. Wu Jiao and Du Juan, "Work Starts on Major Stage of Gas Pipeline," *China Daily*, September 15, 2014, http://usa.chinadaily.com.cn/epaper/2014-09/15/content_18600154.htm.

12. John A. Mathews and Hao Tan, "The Greening of China's Black Electric Power System? Insights from 2014 Data," *Asia-Pacific Journal* 13, issue 10, no. 2 (2015), http://www.japanfocus.org/-John_A_-Mathews/4297/article.html.

13. Vernon Henderson, Todd Lee, and Yung Joon Lee, "Scale Externalities in Korea," *Journal of Urban Economics* 49, no. 3 (2001): 479–504.

14. Sylvie Démurger, Jeffrey D. Sachs, Wing Thye Woo, Shuming Bao, Gene Chang, and Andrew Mellinger, "Geography, Economic Policy, and Regional Development in China," *Asian Economic Papers* 1, no. 1 (2002): 146–97.

15. National Bureau of Statistics of China, *China Energy Statistical Yearbook 2006* (Beijing: China Statistics Press, 2006).

16. J. C. Witte, M. R. Schoeberl, A. R. Douglass, J. F. Gleason, N. A. Krotkov, J. C. Gille, K. E. Pickering, and N. Livesey, "Satellite Observations of Changes in Air Quality during the 2008 Beijing Olympics and Paralympics," *Geophysical Research Letters* 36, no. 17 (2009), http://onlinelibrary.wiley.com/doi/10.1029/2009GL039236/full.

17. Siqi Zheng, Cong Sun, Ye Qi, and Matthew E. Kahn, "The Evolving Geography of China's Industrial Production: Implications for Pollution Dynamics and Urban Quality of Life," *Journal of Economic Surveys* 28, no. 4 (2014): 709–24.

18. "Shougang Quits Blast Furnace for Olympics," *China Daily*, January 6, 2008, http://www.chinadaily.com.cn/olympics/2008-01/06/content_6373382.htm.

19. Qu Yue, Fang Cai, and Xiaobo Zhang, "Has the 'Flying Geese' Phenomenon in Industrial Transformation Occurred in China?" In *Rebalancing and Sustaining Growth in China*, ed. Huw McKay and Ligang Song, 93–110 (Acton, Australian Capital Territory, Australia: ANU E Press, 2012).

20. Matthew E. Kahn, "The Silver Lining of Rust Belt Manufacturing Decline," *Journal of Urban Economics* 46, no. 3 (1999): 360–76.

21. Siqi Zheng, Jing Cao, Matthew E. Kahn, and Cong Sun, "Real Estate Valuation and Cross-Boundary Air Pollution Externalities: Evidence from Chinese Cities," *Journal of Real Estate Finance and Economics* 48, no. 3 (2014): 398–414.

22. Zheng et al., "The Evolving Geography."

23. "The Post-Industrial Future Is Nigh," *Economist*, February 19 2013, http://www.economist.com/blogs/analects/2013/02/services-sector; National Energy Administration, "The Forecast of China's Medium- to Long-Term Power Generation Capacity and Power Demand," February 20, 2013, http://

www.nea.gov.cn/2013-02/20/c_132180424_4.htm (in Chinese). For more evidence about China's 65redistribution of manufacturing across regions, see Zheng et al., "The Evolving Geography."

24. Kenneth Y. Chay and Michael Greenstone, *The Impact of Air Pollution on Infant Mortality: Evidence from Geographic Variation in Pollution Shocks Induced by a Recession*, NBER Working Paper 7442 Cambridge MA: National bureau of economic research, 1999.

25. Jian Xie and Fasheng Li, *Overview of the Current Situation on Brownfield Remediation and Redevelopment in China* (Washington, DC: World Bank, 2010).

26. George Akerlof, "The Market for 'Lemons': Quality Uncertainty and the Market Mechanism," *Quarterly Journal of Economics* 84, no. 3 (1970): 488–500. It should be noted that, more recently, used car sites such as CARFAX (http://www.carfax.com) have brought much needed information to the buying public.

27. Hong Kong Environment Bureau, *A Clean Air Plan For Hong Kong*, http://www.enb.gov.hk/en/files/New_Air_Plan_en.pdf.

28. Hepeng Jia, "China Blamed for Mercury on Iconic Mount Fuji," *Chemistry World*, October 18, 2013, http://www.rsc.org/chemistryworld/2013/10/china-blamed-mercury-pollution-mount-fuji-japan.

29. Claire Topal and Yeasol Chung, "China's Off-the-Chart Air Pollution: Why It Matters (and Not Only to the Chinese): An Interview with Daniel K. Gardner," January 27, 2014, http://www.nbr.org/research/activity.aspx?id=397#footnote2.

30. The National Quality Standards for Surface Water are released by the Ministry of the Environment and are applicable to the surface water bodies of rivers, lakes, and reservoirs in China. There are five classes in the standards: Class I indicates the highest quality and Class V indicates the worst. See The National Standards of the People's Republic of China, "Environmental Quality Standards for Surface Water," http://english.mep.gov.cn/SOE/soechina1997/water/standard.htm.

31. Hilary Sigman, *International Spillovers and Water Quality in Rivers: Do Countries Free Ride?* NBER Working Paper 8585 (Cambridge, MA: National Bureau of Economic Research, 2001).

32. Cai Hongbin, Yuyu Chen, and Qing Gong, "Polluting Thy Neighbor: The Case of River Pollution in China," working paper, Peking University, 2012.

33. "China's Rural Pollution Problem," *Wall Street Journal*, July 27, 2013, http://online.wsj.com/article/SB10001424127887323382910457862401064822814 2.html?mod=WSJ_hpp_RIGHTTopNewsCollection.

34. "Apparent Win for Shell in Nigeria Spill Cases," *Wall Street Journal*, January 30, 2013, http://online.wsj.com/news/articles/SB10001424127887323829104578624010648228142.

35. Gwynn Guilford, "Half of the Rice in Guangzhou Is Polluted," *Atlantic*, May 21, 2013, http://www.theatlantic.com/china/archive/2013/05/half-of-the-rice-in-guangzhou-is-polluted/276098/.

36. "Rare-Earth Mining in China Comes at a Heavy Cost for Local Villages," *Guardian*, August 7, 2012, http://www.theguardian.com/environment/2012/aug/07/china-rare-earth-village-pollution.

37. Nicholas Bloom, Christos Genakos, Ralf Martin, and Raffaella Sadun, "Modern Management: Good for the Environment or Just Hot Air?" *Economic Journal* 120, no. 544 (2010): 551–72.

38. Zheng Siqi, Matthew E. Kahn, and Hongyu Liu, "Towards a System of Open Cities in China: Home prices, FDI Flows and Air Quality in 35 Major Cities," *Regional Science and Urban Economics* 40, no. 1 (2010): 1–10.

39. Information Technology and Innovation Foundation, "China Has Increased Its Solar PV Global Export Market Share from 2 Percent in 2000 to 54 Percent by 2011, Even Though US Solar PV Are 5 Percent More Cost Competitive Than Chinese Products before Subsidies," http://www.itif.org/content/china-has-increased-its-solar-pv-global-export-market-share-2-2000-54-2011-even-though-us-so.

40. Walmart, "Renewable Energy," http://corporate.walmart.com/global-responsibility/environment-sustainability/energy.

41. Aparna Sawhney and Matthew E. Kahn, "Understanding Cross-National Trends in High-Tech Renewable Power Equipment Exports to the United States," *Energy Policy* 46 (2012): 308–18.

42. "Downward Trend in Global Solar PV Module Prices to Continue through 2015," http://www.marketwired.com/press-release/downward-trend-in-global-solar-pv-module-prices-to-continue-through-2015-1402723.htm.

43. Richard B. Freeman and Wei Huang, *China's "Great Leap Forward" in Science and Engineering*. NBER Working Paper 21081 (Cambridge, MA: National Bureau of Economic Research, 2015).

Chapter 3: The Migration to Cities

1. Lucy Hornby and Jane Lanhee Lee, "In China's Urbanization, Worries of a Housing Shortage," *New York Times*, March 31, 2013, http://www.nytimes.com/2013/04/01/business/global/in-chinas-urbanization-worries-of-a-housing-shortage.html?pagewanted=all&_r=0.

2. "Left-Behind Children of China's Migrant Workers Bear Grown-Up Burdens," *Wall Street Journal*, January 17, 2014, http://online.wsj.com /news/articles/SB10001424052702304173704579260900849637692.

3. Leo Feler and J. Vernon Henderson, "Exclusionary Policies in Urban Development: Under-Servicing Migrant Households in Brazilian Cities," *Journal of Urban Economics* 69, no. 3 (2011): 253–72.

4. Government Office for Science, "Migration and Global Environmental Change," http://www.bis.gov.uk/foresight/our-work/projects/published -projects/global-migration/reports-publications.

5. Warren Karlenzig, letter to the editor, *New York Times*, July 27, 2015, http:// www.nytimes.com/2015/07/27/opinion/big-plans-a-chinese-megacity-pop -130-million.html?ref=opinion&_r=0.

6. Lucas W. Davis and Paul J. Gertler, "Contribution of Air Conditioning Adoption to Future Energy Use under Global Warming," *Proceedings of the National Academy of Sciences* 112, no. 19 (2015): 5962–67.

7. These thirty-five major cities represent all municipalities directly under the federal government, provincial capital cities, and quasi-provincial capital cities in China, and they account for one-fourth of the total urban population in more than six hundred cities. For this subset we have access to a high-quality, transaction-based hedonic home price index by city and quarter. The construction of this city housing price index is based on the real transaction prices of all newly constructed housing units in a city. The municipal housing authority keeps all the transaction contracts of these units in a database. The contract contains the information on the transaction price (yuan per square meter), the dwelling's physical attributes (unit size, floor number, building structure type, decoration status, etc.), and its detailed address, from which locational attributes (distance to the city center, distance to the closest subway stop, etc.) can be derived. A standard hedonic model is used to compute the quarterly price index using all the transaction observations. Every municipal housing authority then reports the index to the state's Ministry of Housing and Urban-Rural Development. This set of hedonic housing price indexes is proprietary data and has not been publicly published. Coauthor Siqi Zheng is on this housing price index team.

8. Wei Shang-Jin and Xiaobo Zhang, *Sex Ratios, Entrepreneurship, and Economic Growth in the People's Republic of China*, NBER Working Paper 16800 (Cambridge, MA: National Bureau of Economic Research, 2011).

9. "Problems of Place: Do Quotas in China's College Admissions System Reinforce Existing Inequalities?" http://www.tealeafnation.com/2013/06

/problems-of-place-do-quotas-in-chinas-college-admissions-system-reinforce
-existing-inequalities/.

10. Lü Minghe, "Shanghai's Dead Pig Story Stretches Back Upstream," *Guardian*,
    March 25, 2013, http://www.theguardian.com/environment/2013/mar/25
    /shanghai-dead-pig-story-upstream.

11. Wang Yanlin, "Better Quality of Life for City's Residents in 2013," *Shanghai
    Daily*, January 28, 2013, http://www.shanghaidaily.com/metro/Better-quality
    -of-life-for-citys-residents-in-2013/shdaily.shtml.

12. For some basic information about the ZJ High-Tech Park, see Wikipedia,
    "Zhangjiang Hi-Tech Park," https://en.wikipedia.org/wiki/Zhangjiang_Hi
    -Tech_Park.

13. Michael E. Porter, "Clusters and the New Economics of Competition," *Harvard Business Review* 76, no. 6 (1998): 77–90.

14. "Universiade Boosts Quality of Life in Shenzhen," *China Daily Forum*,
    August 11, 2011, http://bbs.chinadaily.com.cn/thread-709516-1-1.html.

15. David Brooks, "The Great Migration," *New York Times*, January 24, 2013,
    http://www.nytimes.com/2013/01/25/opinion/brooks-the-great-migration
    .html.

16. Annemarie Schneider, Chaoyi Changa, and Kurt Paulsen, "The Changing
    Spatial Form of Cities in Western China," *Landscape and Urban Planning* 135
    (2015): 40–61.

17. Wikipedia, "Jing-Jin-Ji," http://en.wikipedia.org/wiki/Jing-Jin-Ji.

18. Ian Johnson, "As Beijing Becomes a Supercity, the Rapid Growth Brings
    Pains," *New York Times*, July 19, 2015, http://www.nytimes.com/2015/07/20
    /world/asia/in-china-a-supercity-rises-around-beijing.html?_r=0.

19. Ian Johnson, "China Aims to Move Beijing Government Out of City's
    Crowded Core," *New York Times*, June 25, 2015, http://www.nytimes.com
    /2015/06/26/world/asia/china-aims-to-move-beijing-government-out-of
    -citys-crowded-core.html?_r=0.

20. Esteban Rossi-Hansberg and Pierre-Daniel Sarte, "Firm Fragmentation
    and Urban Patterns," *International Economic Review* 50, no. 1 (2009):
    143–86.

21. Gulangyu was a place of residence for Westerners during Xiamen's colonial
    past, and is famous for its architecture and for China's only piano museum.
    There are more than two hundred pianos on the island, lending it the nick-
    names Piano Island, The Town of Pianos, or The Island of Music. Its Chinese
    name also has musical roots: *Gulang* (which means "drum waves") refers
    to the sound generated by the ocean waves hitting the reefs, and *yu* means
    "islet." See Wikipedia, "Gulangyu Island," https://en.wikipedia.org/wiki
    /Gulangyu_Island.

22. Based on our survey results, the annual average income is 63,000 yuan in Chengdu, 103,000 yuan in Beijing, and 109,000 yuan in Shanghai and Shenzhen.

23. China Odyssey Tours, "Local Lifestyle of Chengdu—The Most Laid-back Lifestyle," http://www.chinaodysseytours.com/chengdu/local-lifestyle-of -chengdu.html.

24. The secondary industry accounted for 48 percent of the city's total GDP in 2010. Hong Kong Trade Development Council, "Lanzhou (Gansu) City Information," http://china-trade-research.hktdc.com/business-news/article /Fast-Facts/Lanzhou-Gansu-City-Information/ff/en/1/1X000000/1X09RCR9 .htm.

25. W. Walker Hanlon and Yuan Tian, "Killer Cities: Past and Present," *American Economic Review* 105, no. 5 (2014): 570–75.

26. Jonathan Kaiman, "China to Flatten 700 Mountains for New Metropolis in the Desert," *Guardian*, December 6, 2012, http://www.theguardian.com/world /2012/dec/06/china-flatten-mountain-lanzhou-new-area.

27. UK Foreign and Commonwealth Office, *China: Central Government Vision on Hukou Reform*, https://www.gov.uk/government/publications/hukou-reform -central-government-sets-out-a-vision-august-2014/hukou-reform-central -government-sets-out-a-vision-august-2014.

28. Ankit Panda, "China Announces Limited Hukou Reform," *Diplomat*, July 31, 2014, http://thediplomat.com/2014/07/china-announces-limited-hukou -reform/.

29. Li Keqiang, "Address by Li Keqiang: English," https://www.iiss.org/en/events /events/archive/2014-0f13/june-d70b/pm-li-166a/english-c9d4.

30. Media stories report that after the opening of the Beijing–Tianjin bullet train, some large companies sent their manufacturing sections to Tianjin while keeping their headquarters in Beijing. Some information technology engineers bought their homes in Tianjin for less-expensive housing prices. They work at home and commute by bullet train once a week to Beijing to have meetings at their company's headquarters in Beijing.

31. Siqi Zheng and Matthew E. Kahn, "China's Bullet Trains Facilitate Market Integration and Mitigate the Cost of Megacity Growth," *Proceedings of the National Academy of Sciences* 110, no. 14 (2013): E1248–53.

Chapter 4: The Causes and Consequences of Chinese Suburbanization

1. Siqi Zheng, Rui Wang, Edward L. Glaeser, and Matthew E. Kahn, "The Greenness of China: Household Carbon Dioxide Emissions and Urban Development," *Journal of Economic Geography* 11, no. 5 (2010): 761–92.

2. Ming Lu, Cong Sun, and Siqi Zheng, *Congestion and Pollution Consequences of Driving-to-School Trips: A Case Study in Beijing*, working paper, Hang Lung Center for Real Estate, Tsinghua University, 2015.

3. Edward L. Glaeser, Jed Kolko, and Albert Saiz, "Consumer City," *Journal of Economic Geography* 1, no. 1 (2001): 27–50.

4. Joel Waldfogel, "The Median Voter and the Median Consumer: Local Private Goods and Population Composition," *Journal of Urban Economics* 63, no. 2 (2008): 567–82.

5. Zheng Siqi and Matthew E. Kahn, "Does Government Investment in Local Public Goods Spur Gentrification? Evidence from Beijing," *Real Estate Economics* 41, no. 1 (2013): 1–28.

6. Glaeser et al., "Consumer City."

7. Li Yang, "Transform Shantytowns to Improve People's Lives," *China Daily*, July 4, 2014, http://www.chinadaily.com.cn/opinion/2014-07/04/content _17649053.htm.

8. The lease terms are seventy years for residential use, forty years for commercial use, fifty years for industrial and institutional use, and fifty years for mixed use. There is some uncertainty concerning whether the Communist Party will extend the land lease (with additional fee) or reclaim the land when the lease ends.

9. Siqi Zheng, Weizeng Sun, Jianfeng Wu, and Matthew E. Kahn. *The Birth of Edge Cities in China: Measuring the Spillover Effects of Industrial Parks*. NBER Working Paper 21378 (Cambridge, MA: National Bureau of Economic Research, 2015).

10. Ibid.

11. Megha Rajagopalan, "China's Wild West," *Foreign Policy*, November 23, 2011, http://www.foreignpolicy.com/articles/2011/11/23/china_s_wild_west.

12. Chengri Ding and Yan Song, eds. Emerging Land and Housing Markets in China (Cambridge, MA: Lincoln Institute of Land Policy, 2005).

13. The joint report by the World Bank and the Development Research Center of China's State Council, *Urban China: Toward Efficient, Inclusive and Sustainable Urbanization*, http://www.worldbank.org/en/country/china/publication /urban-china-toward-efficient-inclusive-sustainable-urbanization.

14. National Bureau of Statistics of China, *China Statistical Yearbook 2009* (Beijing: China Statistics Press, 2009); National Bureau of Statistics of China, *China Statistical Yearbook 2011* (Beijing: China Statistics Press, 2011).

15. Andrew Jacobs, "Farmers in China's South Riot over Seizure of Land," *New York Times*, September 23, 2011, http://www.nytimes.com/2011/09/24 /world/asia/land-dispute-stirs-riots-in-southern-china.html?ref=world.

16. Nathaniel Baum-Snow, "Did Highways Cause Suburbanization?" *Quarterly Journal of Economics* 122, no. 2) (2007): 775–805.

17. Nathaniel Baum-Snow, Loren Brandt, J. Vernon Henderson, Matthew A. Turner, and Qinghua Zhang, "Roads, Railways and Decentralization of Chinese Cities," unpublished manuscript, 2015.

18. Rui Wang and Quan Yuan, "Are Compact Cities Greener? Evidence from China, 2000–2010," working paper, Ziman Center for Real Estate, University of California–Los Angeles, 2014.

19. Bullitt Center, "Building Features," http://www.bullittcenter.dreamhosters .com/building/building-features/.

20. "95 Pct of New Buildings in China Energy-Inefficient: Official," *Xinhuanet*, November 20, 2011, http://news.xinhuanet.com/english2010/china/2011 -11/20/c_131258646.htm.

21. Siqi Zheng, Jing Wu, Matthew E. Kahn, and Yongheng Deng, "The Nascent Market for 'Green' Real Estate in Beijing," *European Economic Review* 56, no. 5 (2012): 974–84.

22. Samuel R. Dastrup, Joshua Graff Zivin, Dora L. Costa, and Matthew E. Kahn, "Understanding the Solar Home Price Premium: Electricity Generation and 'Green' Social Status," *European Economic Review* 56, no. 5 (2012): 961–73.

23. Li Zhang, Cong Sun, Siqi Zheng, and Hongyu Liu, *The Role of Public Information in Increasing Homebuyers' Willingness-to-Pay for Green Housing: Evidence from Beijing*, working paper, Hang Lung Center for Real Estate, Tsinghua University, 2015.

24. Asian Development Bank, *Eco-City Development: A New and Sustainable Way Forward?*, November 2010, http://www.adb.org/publications/eco-city -development-new-and-sustainable-way-forward.

25. Axel Baeumler, Ede Ijjasz-Vasquez, and Shomik Mehndiratta, eds., *Sustainable Low-Carbon City Development in China* (Washington, DC: World Bank, 2012), http://elibrary.worldbank.org/doi/abs/10.1596/978-0-8213-8987-4.

26. Sue-Lin Wong and Clare Pennington, "Steep Challenges for a Chinese Eco-City," February 13, 2013, http://green.blogs.nytimes.com/2013/02/13/steep -challenges-for-a-chinese-eco-city/.

27. Matthew E. Kahn, "Demographic Change and the Demand for Environmental Regulation," *Journal of Policy Analysis and Management* 21, no. 1 (2002): 45–62.

28. Urban Redevelopment Authority, "Tianjin Eco-city Breaks New Ground," http://www.ura.gov.sg/skyline/skyline08/skyline08-05/text/03.htm.

29. Simon Joss, Daniel Tomozeiu, and Robert Cowley, *Eco-Cities: A Global Survey 2011*, http://www.westminster.ac.uk/?a=119909.

30. Axel Baeumler, Mansha Chen, Arish Dastur, Yabei Zhang, Richard Filewood, Khiary Al-Jamal, Charles Peterson, Monali Ranade, and Nat Pinnoi, *Sino-Singapore Tianjin Eco-City (SSTEC): A Case Study of an Emerging Eco-City in China*, http://www-wds.worldbank.org/external/default/WDSContentServer /WDSP/IB/2011/01/17/000333037_20110117011432/Rendered/PDF/590120 WP0P114811REPORT0FINAL1EN1WEB.pdf.

31. Genia Kostka, *Barriers to the Implementation of Environmental Policies at the Local Level in China*, World Bank Policy Research Working Paper 7016 (Washington, DC: World Bank, 2014).

32. Ibid.

Chapter 5: Private Vehicle Demand in Urban China

1. Research and Innovative Technology Administration, "BTS Publications (Alphabetical List)," http://www.bts.gov/publications/national_transportation _statistics/html/table_01_11.html.

2. Li Shanjun, Junji Xiao, and Yimin Liu, *The Price Evolution in China's Automobile Market*, http://papers.ssrn.com/sol3/papers.cfm?abstract_id=2177749.

3. Ibid.

4. Aaron S. Edlin and Pinar Karaca-Mandic, *The Accident Externality from Driving*, http://ideas.repec.org/a/ucp/jpolec/v114y2006i5p931-955.html.

5. Donald C. Shoup, *The High Cost of Free Parking* (Chicago: APA Planners Press, 2005).

6. Michael Cooper and Jo Craven McGinty, "A Meter So Expensive, It Creates Parking Spots," *New York Times*, March 15, 2012, http://www.nytimes .com/2012/03/16/us/program-aims-to-make-the-streets-of-san-francisco -easier-to-park-on.html.

7. Gabe Collins and Andrew Erickson, "Dying for a Spot: China's Car Ownership Growth Is Driving a National Parking Space Shortage," January 10, 2011, http://www.chinasignpost.com/2011/01/10/dying-for-a-spot-chinas -car-ownership-growth-is-driving-a-national-parking-space-shortage/.

8. Siqi Zheng, *The Spatial Structure of Urban Economy* (Beijing: Tsinghua University Press, 2012), 17–20 (in Chinese).

9. Siqi Zheng, Richard B. Peiser, and Wenzhong Zhang, "The Rise of External Economies in Beijing: Evidence from Intra-Urban Wage Variation," Regional Science and Urban Economics 49 (2009): 449–59.

10. "Beijingers Wasting More Time in Traffic: Study," *People's Daily Online*, December 25, 2009, http://english.peopledaily.com.cn/90001/90782/90872 /6852168.html.

11. Jonathan Leape, "The London Congestion Charge," *Journal of Economic Perspectives* 20, no. 4 (2006): 157–76.

12. Gilles Duranton and Matthew A. Turner, "Urban Growth and Transportation," *Review of Economic Studies* 79, no. 4 (2012): 1407–40.

13. Wikipedia, "Guangzhou Bus Rapid Transit," http://en.wikipedia.org/wiki/Guangzhou_Bus_Rapid_Transit.

14. Xu Jingxi, " Sustainable Development," *China Daily*, November 21, 2012, http://europe.chinadaily.com.cn/travel/2012-11/21/content_15947319.htm.

15. David Cohen, "Enter the People Movers," *China Daily*, April 15, 2011, http://usa.chinadaily.com.cn/business/2011-04/15/content_12333617_2.htm.

16. Shanjun Li, Matthew E. Kahn, and Jerry Nickelsburg, *Public Transit Bus Procurement: The Role of Energy Prices, Regulation and Federal Subsidies*. NBER Working Paper 19964 (Cambridge, MA: National Bureau of Economic Research, 2014).

17. Ian W. H. Parry and Kenneth A. Small, "Does Britain or the United States Have the Right Gasoline Tax?" *American Economic Review* 95, no. 4 (2005): 1276–89.

18. C.-Y. Cynthia Lin and Jieyin Zeng, "The Optimal Gasoline Tax for China," *Theoretical Economics Letters* 4 (2014): 270–78.

19. "Road Pricing in Shanghai: 'Pay Up or Get On Your Bike!' " http://www.shanghaiexpat.com/article/road-pricing-shanghai-pay-or-get-your-bike-552.html.

20. Leape, "The London Congestion Charge"; Kenneth Small, "Road Pricing and Public Transit: Unnoticed Lessons from London," *Access* 26, no. 3 (2005): 10–15.

21. See the discussion in Francisco Posada Sanchez, Anup Bandivadekar, and John German, *Estimated Cost of Emission Reduction Technologies for Light-Duty Vehicles*, http://www.theicct.org/sites/default/files/publications/ICCT_LDVcostsreport_2012.pdf.

22. See the discussion in Haikun Wang, Lixin Fu, and Jun Bi, *Cost and Pollution Emissions from Passenger Cars in China*, http://essi.nju.edu.cn/973proj/upload/2011_3_31/Wang_HK_EP2011(1).pdf.

23. See the discussion in "Automotive Beijing V Emission Standards Will Advance the Implementation of the Country Abolished V Standard," http://news.xinhuanet.com/local/2013-09/18/c_117411167.htm (in Chinese).

24. Matthew E. Kahn, "New Evidence on Trends in Vehicle Emissions," *RAND Journal of Economics* 26, no. 1 (1996): 183–96; Matthew E. Kahn and Joel Schwartz, "Urban Air Pollution Progress despite Sprawl: The 'Greening' of the Vehicle Fleet," *Journal of Urban Economics* 63, no. 3 (2008): 775–87.

25. International Council on Clean Transportation, "Policy Update: China V Gasoline and Diesel Fuel Quality Standards," http://www.theicct.org/sites /default/files/publications/ICCTupdate_ChinaVfuelquality_jan2014.pdf.

26. Edward Wong, "As Pollution Worsens in China, Solutions Succumb to In-fighting," *New York Times*, March 21, 2013, http://www.nytimes.com/2013 /03/22/world/asia/as-chinas-environmental-woes-worsen-infighting -emerges-as-biggest-obstacle.html?pagewanted=all&_r=0.

27. "China to Raise Fuel Standards to Combat Pollution," *Wall Street Journal*, February 7, 2013, http://online.wsj.com/news/articles/SB1000142412788732 4906004578287693562980384.

28. Reuters, "China to Take 5 Million Cars Off the Road," *New York Times*, May 27, 2014, http://www.nytimes.com/2014/05/27/business/china-to-take-5 -million-cars-off-the-road.html.

29. Lucas W. Davis and Matthew E. Kahn, "International Trade in Used Ve-hicles: The Environmental Consequences of NAFTA," *American Economic Journal: Economic Policy* 2, no. 4 (2010): 58–82.

30. Li Shanjun, *Better Lucky Than Rich? Welfare Analysis of Automobile License Allocations in Beijing and Shanghai*, http://papers.ssrn.com/sol3/papers.cfm ?abstract_id=2349865.

31. Daqing Hi-Tech Industrial Development Zone, "Wuhan Introduced Prefer-ential Policies to Encourage the Production of New Energy Vehicles," http:// www.dhp.gov.cn/contents/544/15840.html (in Chinese).

32. Alan Ohnsman, "Tesla Eyes China Green Car Incentives as Shares Reach Record," *Business Week*, February 11, 2014, http://www.businessweek.com /news/2014-02-10/tesla-reaches-record-as-investors-anticipate-foreign-sales -boost.

Chapter 6: The Rising Demand for Blue Skies and Urban Risk Reduction

1. Avraham Ebenstein, Maoyong Fan, Michael Greenstone, Guojun He, Peng Yin, and Maigeng Zhou, "Growth, Pollution, and Life Expectancy: China from 1991–2012," *American Economic Review* 105, no. 5 (2015): 226–31.

2. Jessica Wolpaw Reyes, "Environmental Policy as Social Policy? The Impact of Childhood Lead Exposure on Crime," *BE Journal of Economic Analysis and Policy* 7, no. 1 (2007), http://www3.amherst.edu/~jwreyes/papers/Lead CrimeBEJEAP.pdf.

3. John Donohue and Steven Levitt, *The Impact of Legalized Abortion on Crime*, NBER Working Paper 8004 (Cambridge, MA: National Bureau of Economic Research, 2000).

4. James J. Heckman, "Skill Formation and the Economics of Investing in Disadvantaged Children," *Science* 312, no. 5782 (2006): 1900–1902.

5. Gary S. Becker and H. Gregg Lewis, "Interaction between Quantity and Quality of Children," in *Economics of the Family: Marriage, Children, and Human Capital*, ed. Theodore W. Schultz, 81–90 (Chicago: University of Chicago Press, 1974). The one-child policy is binding in the public sector (governments, public institutions and state-owned enterprises), but less binding for parents who work in the private sector (for private and foreign companies). Anyone who violates this rule must pay a high fine. In addition, the parents will be fired if they are employed in the public sector.

6. "Melamine: China Tainted Baby Formula Scandal" (search results), *New York Times*, n.d., http://topics.nytimes.com/top/reference/timestopics /subjects/m/melamine/index.html.

7. National Bureau of Statistics of China, *Population Census Data of China*, 2000, http://www.stats.gov.cn/english/statisticaldata/yearlydata/YB2000e /index1.htm; National Bureau of Statistics of China, *Population Census Data of China*, 2010, http://www.stats.gov.cn/english/Statisticaldata/CensusData /rkpc2010/indexch.htm.

8. Matthew E. Kahn and John G. Matsusaka, "Demand for Environmental Goods: Evidence from Voting Patterns on California Initiatives," *Journal of Law and Economics* 40 (1997): 137; Matthew E. Kahn, "Demographic Change and the Demand for Environmental Regulation," *Journal of Policy Analysis and Management* 21, no. 1 (2002): 45–62; Matthew J. Holian and Matthew E. Kahn, *Household Demand for Low Carbon Public Policies: Evidence from California*, NBER Working Paper 19965 (Cambridge, MA: National Bureau of Economic Research, 2014).

9. Quan Mu and Junjie Zhang, *Air Pollution and Defensive Expenditures: Evidence from Particulate-Filtering Facemasks*, http://papers.ssrn.com/sol3/papers.cfm ?abstract_id=2518032.

10. Shi Yunhan, "Why Chinese Migrant Workers Are Abandoning the Country's Top Cities," *Atlantic*, April 10 2013, http://www.theatlantic.com/china/archive /2013/04/why-chinese-migrant-workers-are-abandoning-the-countrys-top -cities/274860/.

Chapter 7: Recent Empirical Evidence on the Demand for Lower Pollution Levels

1. Siqi Zheng and Matthew E. Kahn, "Does Government Investment in Local Public Goods Spur Gentrification? Evidence from Beijing," *Real Estate Economics* 41, no. 1 (2013): 1–28.

2. Matthew E. Kahn, "Gentrification Trends in New Transit-Oriented Communities: Evidence from 14 Cities That Expanded and Built Rail Transit Systems," *Real Estate Economics* 35, no. 2 (2007): 155–82.

3. Siqi Zheng, Jing Cao, Matthew E. Kahn, and Cong Sun, "Real Estate Valuation and Cross-Boundary Air Pollution Externalities: Evidence from Chinese Cities." *Journal of Real Estate Finance and Economics* 48, no. 3 (2014): 398–414.

4. Siqi Zheng and Matthew E. Kahn, "China's Bullet Trains Facilitate Market Integration and Mitigate the Cost of Megacity Growth," *Proceedings of the National Academy of Sciences* 110, no. 14 (2013): E1248–53.

5. Quan Mu and Junjie Zhang, *Air Pollution and Defensive Expenditures: Evidence from Particulate-Filtering Facemasks*, http://papers.ssrn.com/sol3/papers.cfm?abstract_id=2518032.

6. World Bank, *Cost of Pollution in China: Economic Estimates of Physical Damages* (Washington, DC: World Bank, 2007).

7. Siqi Zheng, Cong Sun, and Matthew E. Kahn, *Self-Protection Investment Exacerbates Air Pollution Exposure Inequality in Urban China*, NBER Working Paper 21301 (Cambridge, MA: National Bureau of Economic Research, 2015).

8. "Blackest Day," *Economist*, January 14, 2013, http://www.economist.com/blogs/analects/2013/01/beijings-air-pollution.

9. Matthew Neidell, "Information, Avoidance Behavior, and Health: The Effect of Ozone on Asthma Hospitalizations," *Journal of Human Resources* 44, no. 2 (2009): 450–78.

10. Kenneth Y. Chay and Michael Greenstone, *Does Air Quality Matter? Evidence from the Housing Market*, NBER Working Paper 6826 (Cambridge, MA: National Bureau of Economic Research, 1998).

11. Siqi Zheng and Matthew E. Kahn, "Land and Residential Property Markets in a Booming Economy: New Evidence from Beijing," *Journal of Urban Economics* 63, no. 2 (2008): 743–57.

Chapter 8: The Central Government's Increased Desire to Promote Environmental Sustainability

1. Wu Jing, Yongheng Deng, Jun Huang, Randall Morck, and Bernard Yeung, *Incentives and Outcomes: China's Environmental Policy*, NBER Working Paper 18754 (Cambridge, MA: National Bureau of Economic Research, 2013).

2. Gregory C. Chow, *China's Environmental Policy: A Critical Survey* (Princeton, NJ: Princeton University Center for Economic Policy Studies, 2010).

3. Natural Resources Defense Council, "From Copenhagen Accord to Climate Action: Tracking National Commitments to Curb Global Warming," http://www.nrdc.org/international/copenhagenaccords/.

4. "China's Pollution Fight Faces Resistance," *Wall Street Journal*, January 22, 2013, http://online.wsj.com/article/SB1000142412788732330110457825748 4144272650.html.

5. Alex L. Wang, "The Search for Sustainable Legitimacy: Environmental Law and Bureaucracy in China," *Harvard Environmental Law Review* 37 (2013), 376–440.

6. Thomas Heberer and Anja Senz, "Streamlining Local Behaviour through Communication, Incentives and Control: A Case Study of Local Environmental Policies in China," *Journal of Current Chinese Affairs* 40, no. 3 (2011): 77–112.

7. Ran Ran, "Perverse Incentive Structure and Policy Implementation Gap in China's Local Environmental Politics," *Journal of Environmental Policy and Planning* 15, no. 1 (2013): 17–39.

8. Alex L. Wang, "The Search for Sustainable Legitimacy: Environmental Law and Bureaucracy in China," *Harvard Environmental Law Review* 37, no. 2 (2013): 365–440.

9. F. G. Hilton and Arik Levinson, "Factoring the Environmental Kuznets Curve: Evidence from Automotive Lead Emissions," *Journal of Environmental Economics and Management* 35, no. 2 (1998): 126–41.

10. Robert T. Deacon, "Public Good Provision under Dictatorship and Democracy." *Public Choice* 139, nos. 1–2 (2009): 241–62; Casey B. Mulligan, Ricard Gil, and Xavier Sala-i-Martin, "Do Democracies Have Different Public Policies than Nondemocracies?" *Journal of Economic Perspectives* 18, no. 1 (2004): 51.

11. Chen Yuyu, Ginger Zhe Jin, Naresh Kumar, and Guang Shi, "The Promise of Beijing: Evaluating the Impact of the 2008 Olympic Games on Air Quality," *Journal of Environmental Economics and Management* 66, no. 3 (2013): 424–43.

12. Qi Ye, Li Ma, Huanbo Zhang, and Huimin Li, "Translating a Global Issue into Local Priority: China's Local Government Response to Climate Change," *Journal of Environment and Development* 17, no. 4 (2008): 379–400.

13. Genia Kostka, *Barriers to the Implementation of Environmental Policies at the Local Level in China*, World Bank Policy Research Working Paper 7016 (Washington, DC: World Bank, 2014).

14. Pierre-François Landry, *Decentralized Authoritarianism in China: The Communist Party's Control of Local Elites in the Post-Mao Era* (New York: Cambridge University Press, 2008), 31.

254 | Notes to Chapter 8

15. Siqi Zheng, Matthew E. Kahn, Weizeng Sun, and Danglun Luo, "Incentives for China's Urban Mayors to Mitigate Pollution Externalities: The Role of the Central Government and Public Environmentalism," *Regional Science and Urban Economics* 47 (2014): 61–71.

16. Wang, "The Search for Sustainable Legitimacy."

17. Kostka, *Barriers*.

18. For details about California's carbon dioxide cap-and-trade pollution permit program, see California Air Resources Board, "Cap-and-Trade Program," http://www.arb.ca.gov/cc/capandtrade/capandtrade.htm.

19. James B. Bushnell, Howard Chong, and Erin T. Mansur, "Profiting from Regulation: Evidence from the European Carbon Market," *American Economic Journal: Economic Policy* 5, no. 4 (2013): 78–106.

20. Matthew E. Kahn, Pei Li, and Daxuan Zhao, *Pollution Control Effort at China's River Borders: When Does Free Riding Cease?* NBER Working Paper 19620 (Cambridge, MA: National Bureau of Economic Research, 2013).

21. Edward Wong, "As Pollution Worsens in China, Solutions Succumb to In-fighting," *New York Times*, March 21, 2013, http://www.nytimes.com /2013/03/22/world/asia/as-chinas-environmental-woes-worsen-infighting -emerges-as-biggest-obstacle.html.

22. Li Jing, "Chief of Prestigious Chinese University 'To Be Appointed Country's Environment Minister,'" *South China Morning Post*, January 27, 2015, http:// www.scmp.com/news/china/article/1692853/tsinghua-chief-be-environment -minister.

23. Chen's papers written in English are listed at Google Scholar, https://scholar .google.com/scholar?hl=en&q=Jining+Chen++engineering&btnG=&as _sdt=1%2C5&as_sdtp=.

24. Matthew E. Kahn, "Environmental Disasters as Risk Regulation Catalysts? The role of Bhopal, Chernobyl, Exxon Valdez, Love Canal, and Three Mile Island in Shaping US Environmental Law," *Journal of Risk and Uncertainty* 35, no. 1 (2007): 17–43.

25. Ruge Gao, "Rise of Environmental NGOs in China: Official Ambivalence and Contested Messages," *Journal of Political Risk* 1, no. 8 (2013), http://www .jpolrisk.com/rise-of-environmental-ngos-in-china-official-ambivalence-and -contested-messages/.

26. Wanxin Li, Jieyan Liu, and Duoduo Li, "Getting Their Voices Heard: Three Cases of Public Participation in Environmental Protection in China," *Journal of Environmental Management* 98 (2012): 65–72.

27. Benjamin van Rooij and Alex Wang, "China's Pollution Challenge," *New York Times*, May 19, 2014, http://www.nytimes.com/2014/05/20/opinion/chinas -pollution-challenge.html?_r=0.

28. Rachel E. Stern and Jonathan Hassid, "Amplifying Silence: Uncertainty and Control Parables in Contemporary China," *Comparative Political Studies* 45, no. 10 (2012): 1230–54.

29. Anthony J. Spires, "Contingent Symbiosis and Civil Society in an Authoritarian State: Understanding the Survival of China's Grassroots NGOs," *American Journal of Sociology* 117, no. 1 (2011): 1–45.

30. Li et al., "Getting Their Voices Heard."

31. Steven Jiang, Will Ripley, and Michael Pearson, "Death Toll in Tianjin Explosions Reaches 112; More Than 90 Still Missing," http://www.cnn.com/2015/08/15/asia/china-tianjin-explosions/.

32. Li et al., "Getting Their Voices Heard," 69.

33. "Explosion Rocks PX Factory in Fujian," *Economic Observer*, August 1, 2013, http://www.eeo.com.cn/ens/2013/0801/247635.shtml.

34. Ho Chi-ping, "Sustainable Strategy on Waste Disposal," *China Daily*, September 6, 2013, http://www.chinadaily.com.cn/hkedition/2013-09/06/content_16948178.htm.

35. Li et al., "Getting Their Voices Heard," 69.

36. Ibid.

37. Ibid.

38. Cai, "Making Room for Public Participation."

39. Yanhua Deng and Guobin Yang, "Pollution and Protest in China: Environmental Mobilization in Context," *China Quarterly* 214 (2013): 321–36.

40. Anna Lora-Wainwright, "The Inadequate Life: Rural Industrial Pollution and Lay Epidemiology in China," *China Quarterly* 214 (2013): 302–20; Deng and Yang, "Pollution and Protest in China."

41. Kostka, *Barriers*.

42. "Inform Public on Environment," *China Daily*, June 12, 2007, http://www.chinadaily.com.cn/opinion/2007-06/12/content_892113.htm.

43. Juan Botero, Alejandro Ponce, and Andrei Shleifer, *Education and the Quality of Government*, NBER Working Paper 18119 (Cambridge, MA: National Bureau of Economic Research, 2012).

44. Benjamin van Rooij, "The People vs. Pollution: Understanding Citizen Action against Pollution in China," *Journal of Contemporary China* 19, no. 63 (2010): 55–77.

45. Luo Shuzhen, "Specification of Environmental Public Interest Case to Effectively Protect the Environment, the Public Interest," http://www.chinacourt.org/article/detail/2015/01/id/1529356.shtml (in Chinese).

46. AFP News, "China Firms Fined Record $26m for Polluting River," December 31, 2014, https://sg.news.yahoo.com/china-firms-fined-record-26m-polluting-river-070229948--finance.html.

47. Lora-Wainwright, "The Inadequate Life."

48. Esther Duflo, Michael Greenstone, and Rema Hanna, "Cooking Stoves, Indoor Air Pollution and Respiratory Health in Rural Orissa," *Economic and Political Weekly* 43, no. 32 (2008): 71–76.

49. Michael Greenstone and Rema Hanna, "Environmental Regulations, Air and Water Pollution, and Infant Mortality in India." *American Economic Review* 104, no. 10 (2014): 3038–72.

50. The petitioners are held in "black jails," which could be anything from a hotel to an empty school, for weeks or even months before being sent home. Austin Ramzy, "New Report Released on China's 'Black Jails,'" *Time*, November 12, 2009, http://www.time.com/time/world/article/0,8599,1938515,00.html.

51. Raymond Fisman and Yongxiang Wang, "The Mortality Cost of Political Connections," *Review of Economic Studies*, forthcoming.

52. Dalia Ghanem and Junjie Zhang, "'Effortless Perfection': Do Chinese Cities Manipulate Air Pollution Data?" *Journal of Environmental Economics and Management* 68, no. 2 (2014): 203–25.

53. Douglas Almond, Yuyu Chen, Michael Greenstone, and Li Hongbin, "Winter Heating or Clean Air? Unintended Impacts of China's Huai River Policy" *American Economic Review* 99, no. 2 (2009): 184–90.

54. "Study Estimates Extent to Which Air Pollution in China Shortens Human Lives," http://phys.org/news/2013-07-extent-air-pollution-china-shortens.html#jCp.

Chapter 9: Will Local Governments Create Green Cities?

1. The names of mayors used in this chapter are aliases.

2. We thank professor Xinye Zheng of Renmin University for this point.

3. Raj Chetty, John N. Friedman, and Jonah E. Rockoff, *The Long-Term Impacts of Teachers: Teacher Value-Added and Student Outcomes in Adulthood*, NBER Working Paper 17699 (Cambridge, MA: National Bureau of Economic Research, 2011).

4. Siqi Zheng, Matthew E. Kahn, Weizeng Sun, and Danglun Luo, "Incentives for China's Urban Mayors to Mitigate Pollution Externalities: The Role of the Central Government and Public Environmentalism," *Regional Science and Urban Economics* 47 (2014): 61–71.

5. Benjamin van Rooij and Alex Wang, "China's Pollution Challenge," *New York Times*, May 19, 2014, http://www.nytimes.com/2014/05/20/opinion/chinas-pollution-challenge.html?ref=opinion&_r=0.

6. Sarah Eaton and Genia Kostka, "Does Cadre Turnover Help or Hinder China's Green Rise? Evidence from Shanxi Province," in *Chinese Environmental Governance: Dynamics, Challenges, and Prospects in a Changing Society*, ed. Bingqiang Ren and Huisheng Shou, 83–111 (New York: Palgrave Macmillan, 2013). Sarah Eaton and Genia Kostka, "Authoritarian Environmentalism Undermined? Local Leaders' Time Horizons and Environmental Policy Implementation in China," *China Quarterly* 218 (2014): 359–80.

7. Michael C. Jensen, and Kevin J. Murphy, "Performance Pay and Top-Management Incentives," *Journal of Political Economy* (1990): 225–64.

8. Chang-Tai Hsieh and Peter J. Klenow, "Misallocation and Manufacturing TFP in China and India," *Quarterly Journal of Economics* 124, no. 4 (2009): 1403–48.

9. Kenneth A. Couch and Dana W. Placzek, "Earnings Losses of Displaced Workers Revisited," *American Economic Review* 100, no. 1 (2010): 572–89.

10. Genia Kostka, *Barriers to the Implementation of Environmental Policies at the Local Level in China*, World Bank Policy Research Working Paper 7016 (Washington, DC: World Bank, 2014).

11. Lei Shen, Tao Dai, and Aaron James Gunson, "Small-Scale Mining in China: Assessing Recent Advances in the Policy And Regulatory Framework," *Resources Policy* 34, no. 3 (2009): 150–57.

12. Eli Berman and Linda T. M. Bui, "Environmental Regulation and Productivity: Evidence from Oil Refineries," *Review of Economics and Statistics* 83, no. 3 (2001): 498–510; J. Vernon Henderson, "The Effect of Air Quality Regulation," *American Economic Review* 86, no. 4 (1996): 789–813; Michael Greenstone, "The Impacts of Environmental Regulation on Industrial Activity," *Journal of Political Economy* 110, no. 6 (2002): 1175–1219; Randy Becker and Vernon Henderson, "Effects of Air Quality Regulations on Polluting Industries," *Journal of Political Economy* 108, no. 2 (2000): 379–421; Matthew E. Kahn and Erin T. Mansur, "Do Local Energy Prices and Regulation Affect the Geographic Concentration Of Employment?" *Journal of Public Economics* 101 (2013): 105–14.

13. Michael E. Porter and Claas Van der Linde, "Toward a New Conception of the Environment-Competitiveness Relationship," *Journal of Economic Perspectives* 9, no. 4 (1995): 97–118.

14. Junguo Liu and Hong Yang, "China Fights against Statistical Corruption," *Science* 325, no. 5941 (2009): 675; Guan Dabo, Zhu Liu, Yong Geng, Sören Lindner, and Klaus Hubacek, "The Gigatonne Gap in China's Carbon Dioxide Inventories," *Nature Climate Change* 2, no. 9 (2012): 672–75.

15. Qiang Zhang, David G. Streets, Kebin He, Yuxuan Wang, Andreas Richter, John P. Burrows, Itsushi Uno, Carey J. Jang, Dan Chen, Zhiliang Yao, and Yu Lei, "NO$_x$ Emission Trends for China, 1995–2004: The View from the Ground and the View from Space," *Journal of Geophysical Research: Atmospheres* 112, no. D22 (2007), doi: 10.1029/2007JD008684.

16. Timothy Besley and Robin Burgess, "The Political Economy of Government Responsiveness: Theory and Evidence from India," *Quarterly Journal of Economics* 117, no. 4 (2002): 1415–51; Claudio Ferraz and Frederico Finan, "Exposing Corrupt Politicians: The Effects of Brazil's Publicly Released Audits on Electoral Outcomes" *Quarterly Journal of Economics* 123, no. 2 (2008): 703–45; Sheoli Pargal and David Wheeler, *Informal Regulation of Industrial Pollution in Developing Countries: Evidence from Indonesia* (Washington, DC: World Bank, 1995); Abhijit Banerjee, Selvan Kumar, Rohini Pande, and Felix Su, "Do Informed Voters Make Better Choices? Experimental Evidence from Urban India," unpublished manuscript, http://www.poverty actionlab.org/node/2764.

17. David Lague, "China Blames Oil Firm for Chemical Spill," *New York Times*, November 25, 2005, http://www.nytimes.com/2005/11/24/world/asia/24 iht-harbin.html?pagewanted=all.

18. BBC News, "China's Rising People Power," http://news.bbc.co.uk/2/hi/asia -pacific/7195434.stm.

19. BBC News, "China Protest Closes Toxic Chemical Plant in Dalian," http:// www.bbc.co.uk/news/world-asia-pacific-14520438.

20. Keith Bradsher, "Bolder Protests against Pollution Win Project's Defeat in China," *New York Times*, July 4, 2012, http://www.nytimes.com/2012/07 /05/world/asia/chinese-officials-cancel-plant-project-amid-protests .html?_r=0.

21. Jane Perlez, "Waste Project Is Abandoned Following Protests in China," *New York Times*, July 28, 2012, http://www.nytimes.com/2012/07/29/world/asia /after-protests-in-qidong-china-plans-for-water-discharge-plant-are -abandoned.html.

22. Tong Zhifeng, "Reflections on Environmental Mass Incidents in China," http://www.brill.com/sites/default/files/ftp/samplechapter/27609-Sample -Article-2007(vol3).pdf.

23. Google Insights is a publicly available online tool for tracking aggregate Internet search intensity over time for specific geographic areas. Baidu is a local search engine that is widely used in China. It started to provide a similar search intensity index in 2006, but we do not use the Baidu index for two

reasons: first, the Baidu index does not cover our study period; second, some have claimed that the Baidu search engine manipulates the relative sorting order of some search outcomes.

24. These key monitored firms include large firms in heavy pollution-producing industries—firms that have had large-scale environmental incidents and large sewage treatment plants.

25. In 2010 Chongqing and Shanghai launched the pilot program for property tax. The newly designed tax is more like a luxury tax, under which only high-end properties (less than 5 percent of the total housing stock) will be taxed. There is no property tax in other cities.

26. Lixing Li, "The Incentive Role of Creating 'Cities' in China," *China Economic Review* 22, no. 1 (2011): 172–81.

27. Bahl Roy, "Fiscal Decentralization as Development Policy," *Public Budgeting and Finance* 19, no. 2 (1999): 59–75. Zhou Ye'an and Zhang Quan, "Financial Decentralization, Economic Growth and Fluctuations," *Management World* 3 (2008): 6–15.

28. Antung A. Liu and Junjie Zhang, "Fiscal Federalism and Privatization: The Case of Sewage Treatment in China," unpublished manuscript, November 1, 2010, http://watereconomics.ucsd.edu/pdfs/Liu_and_Zhang_2010w.pdf.

29. Tao Ran, Fei Yuan, and Guangzhong Cao, "Regional Competition, Land Leasing and Local Fiscal Effect: Analysis Based on the Panel Data in Chinese Level Cities in 1999–2003," *Journal of World Economy* 10 (2007): 15–27 (in Chinese).

30. Hongbin Cai, J. Vernon Henderson, and Qinghua Zhang, "China's Land Market Auctions: Evidence of Corruption?" *RAND Journal of Economics* 44, no. 3 (2013): 488–521. Such land leasehold rights provide the purchaser with seventy years for residential use, forty years for commercial use, and fifty years for industrial use. After the year 2004, leaseholds are, in principle, all sold through public auction.

31. According to the statistics from China's National Audit Office, as of the end of 2010 the total balance of municipality loans was 10.7 trillion yuan for all cities. The number for Beijing is 374.5 billion yuan, accounting for 27.2 percent of Beijing's GDP in 2010.

32. George E. Peterson, *Land Leasing and Land Sale as an Infrastructure-Financing Option*, World Bank Policy Research Working Paper 4043 (Washington, DC: World Bank, 2006).

33. Keith Bradsher, "China Plans Audit of Debt Government Has Incurred," *New York Times*, July 28, 2013, http://www.nytimes.com/2013/07/29/business

/global/broad-audit-of-chinese-government-agencies-set.html?pagewanted =print.

34. Arnott Richard, "Housing Policy in Developing Countries: The Importance of the Informal Economy," in *Urbanization and Growth*, ed. Michael Spence, Patricia Clarke Annez, and Robert M. Buckley (Washington, DC: World Bank Publications, 2009), 167.

35. Tao Ran and Wang Hui, "China's Unfinished Land System Reform: Challenges and Solutions," http://en.cnki.com.cn/Article_en/CJFDTOTAL -GJPP201002008.htm (in Chinese).

36. One recent study concluded that 60 percent of the air pollution mitigation gains were short-lived for these cities; see Chen Yuyu, Ginger Zhe Jin, Naresh Kumar, and Guang Shi, "The Promise of Beijing: Evaluating the Impact of the 2008 Olympic Games on Air Quality," *Journal of Environmental Economics and Management* 66, no. 3 (2013): 424–43.

Chapter 10: Conclusion

1. Joshua Graff Zivin and Matthew Neidell, "Environment, Health, and Human Capital," *Journal of Economic Literature* 51, no. 3 (2013): 689–730.

2. W. Walker Hanlon and Yuan Tian, "Killer Cities: Past and Present," *American Economic Review* 105, no. 5 (2015): 570–75.

3. Mark L. Egan, Casey B. Mulligan, and Tomas J. Philipson, *Adjusting National Accounting for Health: Is the Business Cycle Countercyclical?*, NBER Working Paper 19058 (Cambridge, MA: National Bureau of Economic Research, 2013).

4. Dora L. Costa and Matthew E. Kahn, "Changes in the Value of Life, 1940–1980," *Journal of Risk and Uncertainty* 29, no. 2 (2004): 159–80.

5. Fergus Green and Nicholas Stern, *China's "New Normal": Structural Change, Better Growth, and Peak Emissions* (London: Grantham Research Institute on Climate and the Environment/Centre for Climate Change Economics and Policy, 2015), 3.

6. Gardiner Harris, "Delhi Wakes Up to an Air Pollution Problem It Cannot Ignore," *New York Times*, February 14, 2015, http://www.nytimes.com/2015 /02/15/world/asia/delhi-wakes-up-to-an-air-pollution-problem-it-cannot -ignore.html.

7. "Meet the Biggest Polluter in China's Most Polluted City," *Wall Street Journal*, September 16, 2014, http://online.wsj.com/article/SB100014240529702039 37904580119960842299020.html.

8. Gene M. Grossman and Alan B. Krueger, "Economic Growth and the Environment," *Quarterly Journal of Economics* 110, no. 2 (1995): 353–77.

9. Siqi Zheng, Matthew E. Kahn, and Hongyu Liu, "Towards a System of Open Cities in China: Home Prices, FDI Flows and Air Quality in 35 Major Cities," *Regional Science and Urban Economics* 40, no. 1 (2010): 1–10; Siqi Zheng, Matthew E. Kahn, Weizeng Sun, and Danglun Luo, "Incentives for China's Urban Mayors to Mitigate Pollution Externalities: The Role of the Central Government and Public Environmentalism," *Regional Science and Urban Economics* 47 (2014): 61–71.

10. Jared Diamond, "What's Your Consumption Factor?," *New York Times*, January 2, 2008, http://www.nytimes.com/2008/01/02/opinion/02diamond.html?.

11. John A. Mathews and Hao Tan, "The Greening of China's Black Electric Power System? Insights from 2014 Data," *Asia-Pacific Journal* 13, issue 10, no. 2 (2015), http://www.japanfocus.org/-John_A_-Mathews/4297/article.html.

12. Ian Parry, Chandara Veung, and Dirk Heine, "How Much Carbon Pricing Is in Countries' Own Interests? The Critical Role of Co-Benefits," IMF Working Paper, https://www.imf.org/external/pubs/ft/wp/2014/wp14174.pdf.

13. Costa and Kahn, "Changes in the Value of Life."

14. Daron Acemoglu, Philippe Aghion, Leonardo Bursztyn, and David Hemous, "The Environment and Directed Technical Change," *American Economic Review* 102, no. 1 (2012): 131–66.

15. William Nordhaus, "Climate Clubs: Overcoming Free-Riding in International Climate Policy," *American Economic Review* 105, no. 4 (2015): 1339–70.

16. BBC News, "US and China Leaders in 'Historic' Greenhouse Gas Emissions Pledge," November 12, 2014, http://www.bbc.com/news/world-asia-china-30015545.

17. Lenore Taylor and Tania Branigan, "US and China Strike Deal on Carbon Cuts in Push for Global Climate Change Pact," *Guardian*, November12, 2014, http://www.theguardian.com/environment/2014/nov/12/china-and-us-make-carbon-pledge.

18. Reuters, "China Plan a Market for Carbon Permits," *New York Times*, August 31, 2014, http://www.nytimes.com/2014/09/01/business/international/china-plans-a-market-for-carbon-permits.html.

19. Edmund G. Brown Jr., "Governor Brown Sworn In, Delivers Inaugural Address," January 5, 2015, http://gov.ca.gov/news.php?id=18828.

20. Walmart, "Energy," http://corporate.walmart.com/global-responsibility/environment-sustainability/energy.

21. Michael I. Cragg, Yuyu Zhou, Kevin Gurney, and Matthew E. Kahn, "Carbon Geography: The Political Economy of Congressional Support for Legislation Intended to Mitigate Greenhouse Gas Production," *Economic Inquiry* 51, no. 2 (2013): 1640–50; Matthew J. Holian and Matthew E. Kahn, "Household Demand for Low Carbon Public Policies: Evidence from California," *Journal of the Association of Environmental and Resource Economists* 2, no. 2 (2015): 205–34.

22. Magali A. Delmas, Matthew E. Kahn, and Stephen Locke, *Accidental Environmentalists? Californian Demand for Teslas and Solar Panels*, NBER Working Paper 20754 (Cambridge, MA: National Bureau of Economic Research, 2014).

23. Richard A. Easterlin, Robson Morgan, Malgorzata Switek, and Fei Wang, "China's Life Satisfaction, 1990–2010," *Proceedings of the National Academy of Sciences* 109, no. 25 (2012): 9775–80.

CPSIA information can be obtained
at www.ICGtesting.com
Printed in the USA
LVHW091830281019
635582LV00003B/403/P